DIGITAL LETHARGY

DIGITAL LETHARGY

Dispatches from an Age of Disconnection

TUNG-HUI HU

The MIT Press
Cambridge, Massachusetts
London, England

The MIT Press would like to thank the anonymous peer reviewers who provided comments on drafts of this book. The generous work of academic experts is essential for establishing the authority and quality of our publications. We acknowledge with gratitude the contributions of these otherwise uncredited readers.

This book was set in Scala and ScalaSans by New Best-set Typesetters Ltd. Printed and bound in the United States of America.

Library of Congress Cataloging-in-Publication Data is available.

ISBN: 978-0-262-04711-1

10 9 8 7 6 5 4 3 2 1

CONTENTS

INTRODUCTION

FORGET YOU

Maybe you've glanced down at your phone and caught yourself scrolling mindlessly through a list of friends, rather than choosing to connect with any of them. Maybe you suddenly feel you've run out of words in a world where you are free, even expected, to express yourself, and all you can come up with is three letters: "lol." Maybe a job lead arrives but you find it easier to click away the hours at your current gig than think about your future. There's a recalcitrant set of feelings here—of being passive, or wanting to disassociate and be anyone but yourself, or avoiding decisions—that I call *digital lethargy*. Because they go against the sense of agency and liveness that digital platforms produce, and against the permission they tout for users to simply be themselves, these feelings can seem perverse, even self-defeating. An Amazon warehouse worker who gets exploited for low wages during the workday and then goes home, exhausted, to binge-shop on Amazon might be suffering from digital lethargy.[1]

In ancient Greek medicine, lethargy described an illness whose victims forgot who they were as they slipped into a comatose, object-like state (the word comes from Lethe, the mythological river of forgetting). Usually acute and often fatal, it was typically treated by reminding the patient of his identity: "he should be placed in a light room, moderately heated, and from time to time an attempt should be made to rouse him by calling his name into his ear, tickling or pricking him," wrote Caelius Aurelianus, the Numidian physician who collected and translated cures for lethargy into Latin.[2] Today, the disengaged, lethargic user is "treated" by algorithms that prod the user into individuating themselves through a stream of clicks, or by social networks that remind the user of opportunities missed, or by trackers and sensors that convert even the failure to respond—what those who study consumer choice call the "no-choice option"—into another form of data.[3] Lethargy in a digital age is distinguished by the fact that you're always "on" as far as technology is concerned, even if you think you've logged off.

This is because gathering data about how you are "yourself" is the main way that digital companies make money off from consumers: when, for example, Facebook asks, "What's on your mind?" and sells the resulting data it gleans to advertisers. More broadly, however, digital capitalism works by linking a user's active participation to attention, capital, and consumerism,[4] and by turning one's interests into choices that can be mined through big data. This can feel liberating, as there are no longer many standards or norms of behavior that one is compelled to observe on the Western Internet:[5] one can do whatever one wants—as long as one continues to click and choose, that is. Under digital capitalism, "being yourself" is the dominant set of codes for how we understand ourselves and others. It is a form of empowered individualism, where we equate a user account with personhood, and we equate choice with agency. It is a mandate to continually act and communicate that has come to feel like normal life, indeed, like life expressed, participated in, personalized, and lived more fully, where we are simply how we act and what we choose.

Yet the idea that the freedom to be yourself is the best way that people should exist in the world is complicated. For one, this claim is often experienced by the digital underclass as a professional requirement, rather than a choice. Consider a series of promotional tweets for a ride-sharing app: "My advice for any driver is to be positive, be happy and be yourself!"; "My best tip is to be yourself. I treat riders how I would like to be treated and I think they appreciate it!"[6] Job coaches similarly advise white-collar job seekers to develop "culturally-accepted hobbies," such as biking or photography, in order to fashion a better sense of oneself as a personal brand.[7] Clearly, "being yourself" can be exhausting if it is a form of customer service or marketing. Even for those who don't have to continually market themselves, selfhood is not always a site of agency. Whether this looks like burnout from an overload of decisions to make or as a continual if exhausting demand for you to represent your racial or ethnic identity at school or work, selfhood can sometimes be a weight. Often, though, one is only vaguely aware of a feeling of not wanting to be oneself; lethargy is typically experienced as the absence or a blockage of feeling than as a clearly definable emotion.

The demands that digital networks place upon us—smartphones that clamor for attention, an endless stream of emails and texts to answer and pictures to respond to, the leakage of work into the home and into private life—may make this situation feel like a uniquely contemporary phenomenon. In important ways, it is. But the problems of work and selfhood are also old ailments that have gone by several different names in Western culture, such as sloth, fatigue, burnout, idleness, or vagrancy. Looking at these lethargic ailments allows us to track what society has historically valued *in* a self. Compare burnout today, for instance, to what it looked like in premodern and early modern times. When the deadly sin of acedia (sloth) interrupted early Christian monks' contemplations, this "noonday demon" could be seen as a disease that interfered with a spiritual and symbolic project of—if one indulges the anachronism—"realizing the self."[8] As this project became secularized, Europe's aristocratic society began asking its subjects to "know one's place" in the world.[9] Marshaling examples that range from Dürer's engraving *Melencolia I*, in which "the burden of contemplative self-reflexivity . . . literally weighs down on the new scientifically oriented subject," to Shakespeare's introspective

Hamlet, who is plagued by indecision and inaction, cultural historian Anna Schaffner shows that by the early modern period, selfhood had already been well established as a burden.[10] Lethargy, acedia's descendant, was how selfhood's erosion manifested in the body and mind; it referred to the forgetting and abdication of one's own moral and political duties, such as the duty to one's country or to God. Yet even as cultural mores of the time framed lethargy as emasculating and antisocial, lethargy created an in-between space where people might perform or assume alternate identities, for example, by forgetting one's lowly social class.[11]

Our story begins in earnest, however, with the industrial revolution, when lethargy became an illness that affected a growing mass of workers. Diagnosing a widespread overload of information that led to an exhaustion of the nerves, or what he termed "neurasthenia," the German doctor Wilhelm Erb wrote: "Networks . . . have completely transformed conditions in trade and commerce. The night is used for travel, the day for business . . . major political, industrial, and financial crises involve far greater sections of the population."[12] Writing in 1893, Erb was referring to telegraph networks, not digital networks, but his quote would nonetheless not be out of place today: medical experts, psychologists, and other social scientists have blamed networks (and, often, their connection to globalization) for a variety of ailments that range from burnout, overwork, and distractedness to sleeplessness, disengagement, and antisocial behavior. At the time, though, neurasthenia was also used in pseudoscientific efforts to categorize the world's population by its supposed capacity for civilization and self-governance. As medical historian Rafaela Zorzanelli explains, the neurologist William Beard introduced the term to America with a caveat: neurasthenia "mainly affected the brain-workers, whose supply of nervous energy was destroyed by the impositions of industrialized urban life, and high-class girls, with their delicate nervous systems unfit for the demands of life emerging in major cities . . . it afflicted the so-called more advanced ethnic groups such as Jews, Slavs and Anglo-Saxons, but not the blacks and Asian immigrants,"[13] whom Beard considered simply "muscle-workers."

Research into neurasthenia became part of an attempt in the late nineteenth and early twentieth century to turn the worker into a well-oiled machine. In what historian Anson Rabinbach terms the "discovery of fatigue," scientists and doctors, as well as social reformers and philosophers, began to see the worker's body as a "human motor," not unlike the great engines that powered the factories of the industrial revolution.[14] Yet these human motors could run out of energy if overworked and not provided time to regenerate through rest and nourishment. Earlier moral debates over work ethic, Rabinbach argues, shifted into a question of finding and improving labor power, and producing a human motor that was immune to fatigue. Engineer Frederick Winslow Taylor and his disciples in scientific management, for example, timed each individual motion within a job with stopwatches to optimize a worker's movements at work, rather than simply pushing a worker to their physical limit. Fatigue served as a perfect foil for industrial capitalism: it had the power to stymie production lines that were expanding on a dramatic scale. And the efficiency craze that Taylor sparked was not just confined to the assembly line; it also led to books on "scientific dressmaking," optimized kitchen layouts, and efficiency in churches.

We still talk about being fatigued at work, but increasingly use other words, such as burnout, which was first identified in 1974 and accounts for not just fatigue but two other symptoms: a reduced sense of personal accomplishment and depersonalization, that is, detachment from others and from one's work.[15] That gradual shift in language, from fatigue to burnout, points to an important shift in the idea of work. In the late 1950s, the concept of the human motor and, with it, visions of a society without fatigue began to wane; manual work began to leave the center of social life, and membership in trade unions in the United States began to decline. (By the mid-1980s, Britain, France, and most of the remaining high-income OECD countries had followed this trend.) What began to replace it was work that was "immaterial," that is, communicative: even factory workers were no longer seen as subjects implementing orders but instead as active participants tasked with coordinating production, offering feedback, and interfacing with other teams.[16] In his book *La fatigue d'être soi* (roughly, "The Weariness of Being a Self"), sociologist Alain Ehrenberg explains that by the 1970s, these changes led Western society to increasingly idealize a subject who is less a follower of norms or fixed rules than one who takes responsibility and initiative to make their own rules.[17] As its glorification of entrepreneurs suggests, society began to understand the individual as the sovereign master of their own destiny. However, this expectation of endless autonomy and freedom has become so ingrained into our model of selfhood (motto: "It's my choice") that the task of realizing oneself becomes a nagging if often invisible social pressure. Paradoxically, the endless stream of opportunities can result in a feeling of insufficiency, making autonomy, Ehrenberg claims, the root of burnout and depression.[18]

While the World Health Organization considers burnout to be an occupational phenomenon, digital lethargy, its descendant, isn't tied to a workplace; lethargy instead describes the failures of subjectivity in an age where each minute act of self-expression, choice, and autonomy is converted to data and capital. Digital capitalism updates Ehrenberg's model of the entrepreneurial subject by crafting the figure of a sovereign "user" who issues commands to a computer server and chooses which options software should take (even if they ultimately make no choice at all). Yet even as digital media's operations rely on imagining, as media scholar Wendy Chun puts it, "the difference between an empowered user and a couch potato,"[19] the empowered user was more an accident of technology than anything else. For the user was at first an impersonal accounting device, a byproduct of time-sharing technologies in the mid-1960s that divided an expensive computer's time into infinitesimal increments. Like a grandmaster racing between multiple tables of chess, the computer raced between multiple terminals tethered to it, devoting a tiny slice of time to each terminal. This innovation granted a user account the power to command time, to give orders and have them responded to instantly—something we now call interaction. And as the personal computer industry keeps trying to tell us, interaction is the source of agency, even personhood. We can readily spot this pitch in Apple Computer's Orwellian ad "1984," which aired that year during the Super Bowl and features a row of zombie-like viewers staring slack-jawed as Big Brother lectures them on a TV screen. A woman bursts past them and slings a sledgehammer at the TV, destroying it and ushering a new era of free thought (or at least interactive computing). A TV audience may be lethargic, sapped of personhood, and unable to make choices for themselves, but the idealized user of the computer, the ad implies, is free.

By the time Apple launched its personal computer, however, another invention had already begun to change the idea of the user. Client-server computing, which took off in the early 1980s, allowed users to use resources from a computer elsewhere on a network, such as printing or storage. The computer had become a "server," and the relationship with the user became one media scholar Markus Krajewski describes as the relationship between master and domestic servant;[20] these digital servants, given whimsical names such as Majordomo or Jeeves or Oliver, dedicated themselves to waiting for—and waiting on—their new masters. After the commercialization of the Internet, cloud computing in the 2010s further redefined what (and where) a server was by turning servers into streamable services. If communicating with a server was something like fetching water from a specific well or tank, users could now call up a service located "in the cloud" as if turning the tap to get water—no need to know where that water comes from, and no worries that it'll run out. It inaugurated a new economic model known as "everything as a service," where services are pooled and delivered on demand, and where anything can potentially be a service, whether software subscriptions or streaming music sold by the month, jet engines rented by the flying hour,[21] even, as we will see shortly, human intelligence by the batch.

As a result, digital platforms position the user as a master who is surrounded by ambient services that remain alert for commands, such as a voice assistant who lives in the cloud but listens continuously for their orders. To be sure, the user's mastery is partially fictitious; a continuous stream of decision points asking users to respond to phrasing variant A or variant B, to say "ok" or "cancel" is largely for the purposes of gathering marketing data.[22] Nor is it particularly stable: a constant pressure to decide, particularly when an overwhelming number of choices are on offer, can lead to the feeling of lethargy. Nevertheless, this duality between user and service or servant not only underwrites how networked computers work but, this book argues, even begins to define personhood today. If a bot is an agent with limited autonomy, something that waits for action and is primarily acted upon, then, by implication, a human is one who chooses and acts.

But choice, and its implicit link to personhood, is not afforded to all populations in the same ways. The pseudoscientific tropes that were used to classify populations in the nineteenth century still linger with us today, as the artificial metrics of "personality" used by Harvard University's admissions office readily reveal. Created to separate Asian Americans from white applicants, "personality" played to old stereotypes of Asian Americans as technically proficient but expressionless automatons, showing how some populations are assigned less personhood than others.[23] This may seem like an abstract problem, but remember that "everything as a service" model I mentioned earlier? While it may have started with computers, many of the robotic servers and services that exist today are in fact human; platforms increasingly pool workers who rent out their spare time and resources for small, even minuscule jobs. Whether data gathering (LeadGenius), transcription (Rev), "digital janitor" services that clean social media platforms (many, such as CrowdFlower), or simply human intelligence (Mechanical Turk), an army of freelancers across the globe log on and await instructions but typically have little choice in whom they will work for. "Robotic" work takes place in countries like the Philippines and India and Mexico, whose populations are already stereotyped in the West as being hardworking and technically competent people "inherently" fit for manual labor, for being given commands and executing them.

Just as "brain workers" were defined in contrast to "muscle workers" a century ago, companies that make their money from propping up the active user at the expense of the passive server perpetuate what political theorist Cedric Robinson terms *racial capitalism*. As he writes, capitalism is built on the "enlisting of human reserves" via migrant labor and exaggerating differences so that some regional or ethnic groups could be racialized as barbarians, foreigners, or others naturally suited for inferior work.[24] Robinson gives the example of poor mercenaries—"free lancers"—recruited into sixteenth-century European armies, such as Scottish and Welsh fighting for England, or Aragonese and Basque for Spain. Race was invented by Europe's classification of populations through their differences, and this drive to classify persists and returns in today's digital underclass. In his ethnographic study of Amazon warehouse workers in Peterborough, a deindustrialized city in eastern England, for example, geographer Ben Rogaly uses Robinson's theory to show how managers fuel perceived racial and national differences to foster competition. Workers from former British colonies—dubbed "migrants" by their supervisors, despite their British citizenship—are assigned inferior shifts; Polish workers compete with English workers and end up driving up production quotas for everyone. Similar strategies can be readily found throughout the Internet, even when race is not explicitly mentioned: one digital outsourcing company in Iowa (motto: "Outsource to Iowa—not India") uses its heartland, all-American setting to argue that Indian workers simply don't share Western cultural values, while a Philippine outsourcing company promoting itself to the American market argues that Filipinos innately "have a great eye for detail."[25]

To be sure, tech industrialists would counter that they strive to be race-neutral (and gender-neutral)—indeed, that they try to be inclusive of all identities. While acknowledging problems with bias and discrimination, they point to initiatives to hire more diverse staff and to offer a selection of free Internet services ("Internet-lite") to developing countries. Yet inclusivity is, paradoxically, sometimes the problem. Take, for example, the 2016 uproar over Facebook's option for advertisers to exclude ads to Black and Latinx groups. Facebook's response to this criticism was to launch a privacy and transparency initiative that made its internal advertising preferences that capture race as a "multicultural affinity" visible.[26] These responses indicate that the technology industry understands race and gender as identity markers—in other words, data values—that can be chosen or are user preferences. Yet by tagging persons of color with interests in "African-American culture," "Asian culture," or "Latino culture," algorithms contrast them with the default values of whiteness.[27] The inclusion of these "affinities" only reinforces the power a system of classification exerts over those perceived as different: as the poet Édouard Glissant puts it, "I understand your difference . . . I admit you to existence, within my system."[28]

Racial capitalism operates on what we normally think of as racial identities (e.g., Black, white, Asian) but also encompasses other uses of human difference to accumulate capital.[29] It might operate by racializing Muslims; it might operate through nationality or gender.[30] For race itself is a technology for dividing humans, rather than a biological truth; as Alexander Weheliye, a scholar of Black literature and culture, points out, racialization is "a set of sociopolitical processes that discipline humanity into full humans, not-quite-humans, and nonhumans."[31] By looking only for physical markers of identity, then, we miss how racial capitalism operates in a digital environment that largely displaces those markers from sight. I instead argue that technology platforms sell liveliness and personhood by manufacturing differences between active users and passive servers. A user's tenuous monopoly on agency is bolstered, for instance, by the framing of their servers as feminized assistants with names such as Siri or Alexa; by the training that requires customer service agents to remain impassive, like an object, in the face of abuse; and by a globalized framework that teaches users to continually question if their correspondent is "foreign" or "bot."

Yet the seeming distance between human user and robotic server is a clumsy fiction. To begin, users always exist in an intimate relationship with their servers; developers of time-sharing described this uncommonly close relationship as "programming intimacy" before they eventually settled on the word "interactivity" instead.[32] Second, users are also freelancers, and often providers of services, too. Users decide, as free agents, which digital platforms to contract with and agree to terms with—thus the infamous end-user license agreements on apps or websites that can be as byzantine as an employment contract—and, by doing so, produce the social ties and behavioral history that constitute digital personhood today. Even if not explicitly selling time or creating content, a user is expected to use their free time performing the "free labor" that runs the Internet today: uploading files, tagging photos, writing reviews, and other activities. And third, despite their presumed or performed passivity, servers are of course also users of digital products and services, as corporate efforts to market to and capture data from the "bottom of the (data) pyramid" have long recognized.[33]

Thus, any division between users and servers is not only arbitrary, but deliberately produced by digital platforms that sell liveness and agency for one party and turn the other into a robot, and that profit by producing the protocols that mediate between user and service. (As billion-dollar companies such as Uber repeatedly claim to the public and to the courts, they are technology platforms that connect people, rather than employers of, say, drivers; they mediate while simultaneously producing that difference.) For a server, lethargy is the exhaustion of having only a partial claim on selfhood: of needing to "be yourself" for other people, or alternately of having to suppress it; of being what feminist scholars Neda Atanasoski and Kalindi Vora call "human surrogates,"[34] rather than full humans. And yet this is the same problem that afflicts users: a feeling of selfhood as something out-of-reach, burdensome, or even unwanted that trails the feeling of sovereignty like a leaden shadow. Lethargy is a shared burden that leaks across the user/server divide and ensnares both, however unequal that burden may be.

Those divides are ubiquitous. When I was working on a graduate degree in poetry in 2002, a popular source of employment for my classmates was Early English Books Online (EEBO), which scanned and digitized a huge archive of early modern texts from sources such as the British Museum. Because it was then almost impossible for a computer program to recognize the idiosyncratic print of these old books with any degree of accuracy, EEBO employed a bunch of students, not to copy the books themselves, but instead to supervise Filipino workers who served as human optical-character-recognition engines. Each party's identity was invisible to their counterpart on the other side, which was likely a result of the antagonistic nature of the system: my classmates would check each batch that their counterparts submitted for errors, and if the error rate was high enough, they would reject the batch—meaning that the worker on the other side wouldn't get paid. The presumptive qualification for this job was that each budding poet was thought to have a unique competence for how language could be used or, more precisely, "managed."[35]

Though I took an unorthodox track after school—working first as a network engineer, and then, after another round of schooling, as a professor—I kept up with my poetry classmates and peers. Poetry rarely paid the bills, and hustling became increasingly a matter of working at the margins of the digital economy. (Poets are the cockroaches of the artistic sector; the world may end in a nuclear disaster, but someone will still be writing poems about it afterward and eking out a few cents in the process.) Several of them ended up working for "content farms" such as Demand Media, writing texts entirely designed to attract search engine traffic. Enough people would Google "how to make a peanut butter and jelly sandwich" that a content farm would roll out a short article titled "How to Make a Peanut Butter and Jelly Sandwich," with a set of asinine instructions—spread peanut butter on bread, then spread jelly—stretched out to a full web page. These articles—keyword-heavy, requiring the writer to churn out a new article every fifteen minutes or so—were considered one level up from spam, and indeed, in 2011, Google famously tweaked its algorithm to downgrade content farms and other so-called shallow content from its search results, which abruptly ended the content farms' business model. (My friends quickly found other poorly paid digital jobs, like transcribing audio clips and optimizing search engine results.) When I met someone recently who mentioned, matter-of-factly, that her sister back home in Nigeria worked as a spammer, I felt an uncanny bond, for front-line spammers are writers, too. Cranking out perhaps 500 e-mails a day—any responses get routed to their bosses, who do the actual scamming—they spin tall tales crafted to get through algorithmic email filters. Spammers, too, try to stay afloat within and are exploited (albeit far more strenuously and viciously) by a system that runs on clickbait and requires an endless supply of fresh content. My peers in North America were privileged enough to avoid that kind of work, but they were just a few rungs up on the same ladder of disposability.

I say "disposable" because to some of my readers, spam will be so distasteful that the people behind it seem subhuman: for example, a whole host of apps by names such as "robokiller" offer to use voice chatbots to waste the time of—and ultimately eliminate the plague of—spam callers. Like the pirates that ancient and early modern jurists argued were just outside the pale of humanity, and more closely resembled nonhuman predators such as wolves, bots lie outside—and therefore point to—the limits of the "all" that digital media supposedly includes.[36] We have, on one side of the line, the (predominately white, live, human) bodies of users, and on the other, the (racialized, bot-like, imitative, lethargic) bodies of outsiders. And so the link between personhood and agency that digital culture implements is simply an elaborate way of dividing the world into human and nonhuman, subjects and objects. While it may seem natural or even self-evident that humans have sovereign agency and objects don't—that is, the human is one who acts upon the object—this is a logic, I argue, that disproportionately burdens those who have historically been objectified: women, people of color, and people with disabilities.

To this day, looking for job opportunities for my students, I still see advertisements for tech companies hiring "creative writers"—writers to give their artificial intelligence (AI) assistants personalities, who will write jokes for Siri or Alexa—just as they are simultaneously hiring workers in the Global South to teach AI rudimentary qualities of humanity, like how to recognize a laugh. The lethargy that these workers share across the globe, laboring together at the edge of the human, suggests there is more in common between users and servers than we would normally recognize. And so, rather than attempting to cure lethargy, this book argues that lethargic feelings are important for their own sake. Inside a digital economy that runs on choice, it might help us better understand the perspective of those who don't have great choices.

ENDURING THE DIGITAL

How, then, to study a feeling, rather than a concrete piece of technology or a specific population? While probing the workings of digital capitalism, I found a number of investigative projects that exposed a hidden aspect of the data economy, such as ethnographic and participant-observer studies of digital labor, or information scientists that reverse-engineered opaque technological processes. I also found countless critics and thinkers who proposed new tactics to disrupt the domination of companies such as Google or Microsoft. But I also began to notice a handful of artists, performers, and writers who seemed to work against the grain of their peers. Rather than engaging directly in the boisterous debate over technology's impact on society, their works were recessive, self-defeating, even passive. Though those works held up the contemporary moment for closer examination, they were not easily instrumentalized. Instead, I increasingly saw them as offering language and images for describing a nascent feeling that I began to understand as lethargy.

Each chapter in this book is built around a handful of artworks or performances; at times I focus on a single book or film. They offer us a way to attend more carefully to the affective (and thus collective) forms that the state of digital lethargy produces. By thinking with them and at times engaging in direct conversation with their creators, I link social understandings of lethargy—youth disconnection, exhaustion and sleeplessness, the colonial legacy of perceived laziness, the "too-late" of a generation used to precarity rather than the promises of the good life—with the new forms and techniques of digital culture: killing time by scrolling aimlessly, the looped GIF, robot detection algorithms that block someone from surfing a website too fast, scripted protocols for interacting with automated customer service.

Though only some of my subjects describe themselves as digital artists, they all share concerns—about logistics, surveillance, self-tracking, big data, movement, and work—that touch on central debates about how digital capitalism reaches everyday experience. After all, the impact of data algorithms or the optimizations of logistical networks are felt not just online but through the quality of one's sleep and on the shelves of grocery stores. Many of these artists told me that they felt that they were participating in a debate about how to be political in today's age. A few invoked forms of direct action to describe their work, such as a strike or withdrawal, even as I found myself drawn to the parts of their artworks that departed from those preexisting ways of framing political action. Indeed, precisely by performing lethargy and inaction, they reveal the complex relationship between lethargy and political change. Lethargy comes largely out of moments of ordinariness, such as waiting, killing time, impasse, and deferring a decision, and is a result of endurance and carrying on rather than the resolution of crisis and repair. As a word, lethargy is an etymological cousin to latent (the ancient Greek river Lethe is derived from *lanthanesthai*, to forget, and is akin to *lanthanein*, to lie in wait for), and latent potential can crystallize in an instant, or never materialize at all. As a political feeling, lethargy may precipitate resistance, but more commonly, lethargy occurs without any knowledge of a precipitating action or a resolving event; it is a latency that exists despite temporal norms and codes that tell us nothing is happening.

The artists I write about avoid what the photographer Allan Sekula has described as the photojournalistic mode, which fixates on the "decisive moments" of an event—for instance, a demonstrator getting tear gassed or engaging in violence—and creates caricatures of what the photographer's subjects are really like.[37] In contrast, Sekula continues, an "anti-photojournalistic" approach might instead record "the lulls, the waiting and the margins of events." So, for the photographic sequence *Waiting for Tear Gas*, on the 1999 Seattle protests against the World Trade Organization, Sekula created a carousel of slides that continuously repeats and, in the words of curator Stephanie Schwarz, "provides neither an obvious nor a single narrative line. There is no crescendo in Sekula's record. There is no movement from dawn to dusk or from peace to violence . . . the viewer holds out for the denouement, for the single shot, which never arrives."[38] In this extended present, in the interim period, waiting is both torturous and celebratory, affectively charged but not yet resolved into an emotion or action. Waiting is "in the air," as it were, and waiting with other people is as much a political state as an actual confrontation.

Like Sekula, the artists I study eschew spectacular images. Instead of registering familiar horror at the extremes of digital technology (such as drones or surveillance cameras), they explore the lulls and margins that compose life within digital capitalism. However, they go one step beyond Sekula, whose *Waiting for Tear Gas* suggests that even if a dramatic conflict between political protesters and law enforcement doesn't come, or doesn't get photographed, it is nonetheless anticipated.[39] Lethargy does not suggest such a direct line between waiting and political action. It may become the affective infrastructure—the sense of being next to others— necessary for any collective action,[40] or it may simply be something that remains diffuse, ambient, but that nevertheless charges its subjects with potential.[41] Lethargy describes a period before the question "What must be done?" can be fully articulated, though it may not even be asked, let alone enacted and realized. Yet as the artworks I consider show us, periods of lethargy are not necessarily less saturated with liveliness or joy than any other part of life, even though they may be so endlessly elongated that they take up lifetimes.

Lethargic artworks are dispatches from an age of disconnection. They sit with the ambivalence, disappointments, and desires in the present, and with the experiences of those workers, content moderators, and customer service agents who occupy time "in the meantime" (as anthropologist Sarah Sharma puts it), and who, through their labor, allow their wealthier clients to have more lively, engaged experiences.[42] They explore the disappointments of the present day, rather than narratives of futures that are always about to come, whether dystopian or utopian. By attempting to better describe the time that seems dead or otherwise in suspension, they help to develop different vocabulary for conceiving of time in the digital age.

Today's conversations around time almost invariably reference the speed of contemporary life, which then leads to a diagnosis of exhaustion and a prescription of "slowness" to cure it. (This diagnosis has happened before; as Schaffner quips, each age always imagines itself as the most exhausted age.[43]) We are told to go on a digital detox, to slow down, to consume media slowly and mindfully, as if savoring an exquisite, farm-to-table meal.[44] Setting aside questions of access, this fast/slow axis misses a key fact: speed is only a part of the equation. Our experience of these technologies is formed by cultural and design codes that alternately make time *feel* live or dead, present or absent. For instance, users rated Facebook's security scan feature much more trustworthy and thorough after their engineers introduced a short delay before it completed; presumably, that delay made it feel "live" rather than "canned."[45] From another angle, those weighted down with an excess of time often squander it—or so consumerist societies would see it—by tinkering with objects to replace what cannot simply be imported, or by repairing what is broken-down or obsolete; to consumers who are used to throwing away what they consider trash, these forms of laborious ingenuity and invention register only as wasting time.[46] The feelings of fatigue or burnout or boredom that users experience today are not simply a result of speed or societal acceleration, but also come from the dominant narratives of digital capitalism, which map doing and actualization onto liveness and subjectivity, which promise to turn even unproductive moments into productive ones,[47] and which devalue any form of time that cannot resolve itself.

The time of live, always-on digital capitalism is a temporal binding that fastens us into certain subject positions—notably, toward interaction, responsiveness, continuity, and realized potential.[48] Like other temporal binds, which make history a matter of progressing simply from past to future, or make the contentment of domesticity seem full of plenitude, those positions hide lethargic categories of time from view. But lethargic time constitutes most of life under digital capitalism: time that feels unrealized, endless, unresponsive, at an impasse, "dead." Focusing on lethargy prompts this book to ask: What is it like to live inside "dead" time? Who might not want to be more themselves?[49] What is a theory of digital politics that "refuse[s] the tender script of care"?[50]

These methodological shifts can also help us move away from the presumption that everyone wants to be "humanized," a patronizing assumption that saturates much of the reparative, feel-good criticism on digital culture. Let us return to the Filipino workers I described earlier to see what a perspective that considers lethargy might produce. Those workers digitizing scanned books are now part of a larger phenomenon known as microwork, where teams of workers across the globe laboriously tag facial expressions to train an AI, transcribe speech into text, or—in a seedier side of the industry—act as "click farmers" artificially pumping up "likes" for a post a client wishes to promote. Since at least 2010, a stream of news media, documentary films, TV episodes, and scholarship has produced stories meant to evince compassion for people doing repetitive tasks in "soul-crushing," "hellish" conditions; one typical article reads: "For the workers, though, it is miserable work, sitting at screens in dingy rooms facing a blank wall, with windows covered by bars, and sometimes working through the night."[51] The coverage is framed as an exposé into a hidden layer of the Internet, and each new iteration inevitably comes to a single, audience-rousing consensus: "click farms are the new sweatshops."[52]

Scholars and journalists writing about microworkers typically try to humanize them, often by emphasizing the workers' aspirations for better lives—even while positioning them as people who imitate or ape "proper life." In an article in the *New Republic*, for example, we read about a photogenic young Filipina worker who "can do an almost perfect karaoke rendition of Mariah Carey's 'We Belong Together' . . . she dreams of finishing college at the University of Cebu City after she's saved enough money from working for [her employer]. Once she earns a degree in Web design, she'll join the Philippine diaspora and find a job in Australia, New Zealand, or the United States. And during weekends, maybe she can lead a life similar to [Ashley] Nivens,"[53] the (fake) white American student account she creates in her job selling "likes" on social media posts. It is unclear whether the interviewer ever asked if she enjoyed the work, or even entertained the possibility that she may enjoy the work for what it is rather than as a stepping-stone to becoming an American.[54]

The irony is that however menial, repetitive, and unlikable their work may be, many microworkers express satisfaction in how they've managed to find a niche within the Internet's shadow economy. They talk about how digital platforms pull the wool over peoples' eyes, selling fuzzy ideas such as "friendship" and "like" to consumers on trillion-dollar scales; any work they get clicking on "likes" is simply another part of that economy.[55] They describe the work as a way of practicing or building skills, such as learning English or information technology; after all, they tend to be younger and better educated than the general population. Depending on the microwork platform, somewhere between 70 and 85 percent of them have at least a bachelor's degree.[56] And they often see their jobs as a meaningful pathway for surviving a poor economy or waiting out a downturn. Not only do they find ways of enduring the pressures of the digital economy, but they show that endurance itself is a blind spot for critics. Czech playwright Karel Čapek coined the word "robot" in 1920 to describe a laborer who can endure tasks that ordinary humans cannot. In a similar way, Western critics position microworkers as robotic both because of their work—a narrative that instead privileges expressive, skillful, or "meaningful" forms of work, though how that is defined is heavily biased[57]—and also because their endurance seems passive, rather than agentive: they aren't trying to defy their bosses.

Microworkers are fully aware of the innumerable ways that Western-owned digital platforms take advantage of them, but they exist in a more ambivalent relationship with their employers than simply oppression or exploitation. When they express satisfaction in their job, this baffles some scholars of microwork, who suspect them of not understanding their own plight.[58] But that says more about how critics and scholars tend to see the world than anything else. Describing her initial recoil at women participating in the mosque movement in the Middle East—a movement largely overseen by male counterparts—anthropologist Saba Mahmood initially wondered why the women she studied didn't speak out or resist this seeming position of subordination but simply accepted their situation. She soon came to realize that her feelings of discomfort came out of her own sensibility as a secular feminist, and, on reflection, argues that liberal imaginations of freedom can be limiting: "Does the category of resistance impose a teleology of progressive politics on the analytics of power—a teleology that makes it hard for us to see and understand forms of being and action that are not necessarily encapsulated by the narrative of subversion and reinscription of norms?"[59] Key to Mahmood's analysis is the idea that liberal notions of freedom, as action that comes from one's own will, rather than from society or norms, shape how both left-leaning thinkers and also libertarians and neoliberals on the right understand individual agency. In other words, they tend to see agency as action that resists the constraints of power, and miss actions that might be indifferent to or even want to subsist within norms or constraints.

Scholars and critics typically respond to domination by inverting the terms, showing how agency exists in marginalized subjects and in unexpected places and practices, rather than just in (white) men in power. Well-known studies have chronicled the power of foot-dragging,[60] refusal or withdrawal (saying "I prefer not to"),[61] obfuscation and disappearance,[62] disruption,[63] and other forms of "everyday resistance."[64] On the surface, lethargy may sound like a similar topic, and this book is indebted to these critical insights. But as I argue, lethargy both questions the premise of willfulness inherent in most of those approaches and offers ethical and political reasons why we should be willing to be the ones acted upon rather than the ones who act. Instead of using opposition as the default framework for understanding digital capitalism, I endeavor to take those entanglements between worker and platform seriously. Feelings of attachment and aspiration expressed by microworkers are not naïve; instead, they offer a way of reconceptualizing digital culture from the point of view of those who rarely become visible subjects but are more commonly part of its supply chain. In turn, artworks that similarly work within the supply chain—at times complying with or even embracing an algorithm's dictates—help us to pay attention to the self-sabotaging thoughts that spring from the lethargy at the periphery of our digital lives.

This book groups its main questions into two intersecting axes, both of which are contained within the word *lethargos* itself. First, as this section has proposed, lethargy is concerned with enduring a condition rather than refusing it. What on first glance seems like a lack of engagement, action, or responsiveness (*argos*, not-working, not-acting) is in fact a way of abiding, remaining intact, or tolerating the intolerable; it is generally a set of tactics to survive within a condition (e.g., delaying tactics that forestall the inevitable), rather than a way of overturning that condition. It is striking that we almost always hear "abide" in the negative sense—as in a thing we can't abide—rather than its positive sense, of waiting and remaining with; that diminished imagination in the latter sense is something I aim to replenish.

Second, as we saw earlier in this introduction, lethargy is a problem of self-forgetting; it sidesteps the liberal-democratic narrative in which a subject "finds one's voice" and begins to participate in political change. In this vein, I describe artworks that temporarily set aside the burden of constant self-actualization and self-realization—burdens often epitomized by the online user—by exploring the rich if fraught ideas of forgetfulness, imitation, silence, inanimatedness, or even concealment (*lethe*). In Julia Leigh's 2011 film *Sleeping Beauty*, for instance, the main character, a sex worker, drugs herself so that she is asleep and unconscious during her "sleep jobs"; in artist Katherine Behar's choreography, the human elements perform as objects rather than subjects. These attempts to undo the privileged position of the agentive subject can help us understand both the strange status of repetitive and quasi-robotic labor in today's digital age and the fact that this labor is often racialized or gendered, performed not by humans but by "things."

The book's chapters weave ideas from these two threads to make a network. The first two chapters set out the terms for the rest of the book: what lethargy has to do with time, why lethargy requires embracing our objecthood, and why lethargic art tells different kinds of stories about digital capitalism. Chapter 1, "Start When It's Too Late," uses Heike Geissler's memoir of working in an Amazon warehouse, *Seasonal Associate*, to develop a theory of exhausted or "lethargic" agency by examining what's possible when one simply treads water or passes time. Chapter 2, "Wait, Then Give Up," situates lethargic art in the tradition of what art historian Christine Ross terms "depressive aesthetics"—seemingly disengaged art in which the subject is unable to create critical distance from a system of power. Artworks by Cory Arcangel, Katherine Behar, and Tega Brain and Surya Mattu construct the figure of an "unfit" user who fails to use digital culture correctly, but who can nevertheless redirect a viewer's focus to the invisible forms of communicative labor within it.

The following chapters explore the uneven burdens of lethargy, tracking how liveness and interiority are both manufactured and shouldered by today's servers. Chapter 3, "Laugh Out Loud," considers Mexico City–based artist Yoshua Okón's *Canned Laughter* (*Risas enlatadas*) (2009), which depicts a dystopian world where low-wage workers across the US-Mexico border laugh and otherwise emote for white audiences as an allegory for today's systems of microwork, as well as the way those systems continually demand animatedness from racialized workers. Chapter 4 takes up the dissociative performances in Julia Leigh's aforementioned film *Sleeping Beauty* to argue that we might be moving away from the modernist premise of a subject's interiority and toward a cybernetic model of "black box" consciousness. This is not a nihilistic move, however, but an opportunity to reconceive what liveness, privacy, and humanness are today, and, as a result, what we take to be political action by workers who merely endure.

The final chapters explore the new forms of relation that lethargy can offer. While critics often search for more authentic modes of community and engagement on the Internet, lethargy instead starts from the bare fact of bodies and objects idling next to each other. Chapter 5, "Feel Normal," takes up how artist Erica Scourti abdicates her own authenticity and creates a more diffuse form of selfhood that works in concert with, rather than against, social algorithms, such as the autocorrect feature on her iPhone. Chapter 6, "Do Nothing Together," turns to choreographer and dancer nibia pastrana santiago, whose performances help us understand how the very act of doing nothing is foreclosed or curtailed in today's always-on society, particularly for persons of color who have been historically stereotyped as lazy and lacking initiative.

It is common now to argue that digital capitalism is flawed, whether due to concerns about privacy and dataveillance, the way it exacerbates societal inequalities, the monopolistic power of companies, or the bias of its algorithms. It seems we are in a tipping point or crisis induced by technology, and that something must be done. But we won't fully understand it if we focus only on the crisis mentality that the technology itself constructs.[65] Questions of lethargy delay us from diagnosing and stabilizing various forms of being into definable states (human, robot; subject, object) and from prescribing a redemptive form of action. Lethargy is a drag: it weighs down our ability to rush to solutions, and forces us to listen to the unresolved present.

1 START WHEN IT'S TOO LATE

WE APOLOGIZE FOR ANY INCONVENIENCE

In 2016, an autonomous drone bearing a bag of popcorn and a streaming device from Amazon touched down in rural Cambridgeshire, England, reportedly thirteen minutes after an order had been placed. PR images released periodically from its drone research complex underscore Amazon's seemingly magical ability to keep speeding up its delivery times. Amazon was initially focused on building a faster website: an internal study showed that revenue declined by 1 percent for each tenth of a second Amazon's website kept customers waiting.[1] But now Amazon has extended its focus to warehousing and delivery, and has shrunk delivery times from days to overnight to same-day. It uses software to show pickers the fastest route within a warehouse to locate and retrieve their goods, while new robots can carry entire shelves over to their human counterparts, rather than the other way around.[2] These strategies from the world's largest e-commerce company typify the way we commonly imagine digital capitalism: in pursuit of relentless speed and optimization.

It's not surprising, then, that Heike Geissler's book *Seasonal Associate*, a lightly fictionalized memoir of her time working as a temporary employee in an Amazon warehouse in Leipzig, Germany, is filled with references to a culture of speed. There's a routine of sorts: one of Amazon's passive-aggressive team members walks over to give her daily "feedback" about her slow performance or offers tips about how to be more efficient in her work. Lunch breaks are short, so Geissler learns to wolf down her food and then leave a conversation in mid-sentence. Five minutes' tardiness, because of the tram system failing, costs her fifteen minutes of pay; a hipster manager lurks in wait near the exit to catch people who try to clock out too early. What results is, predictably, workers who are always on their feet and told to rush from one point to another—that rush being its own twisted selling point: as one company worker tells new employees during orientation, "it saves him from joining a gym."[3]

Yet the narrative of technology speeding everything up is more complex than it initially appears. Early on, still as a trainee, Geissler observes that there are places where it is indeed possible to be slow at Amazon, where "working time is allowed to melt away . . . outside of the glass container . . . where the more important bosses and planners sit at their desks."[4] Her receiving job in the warehouse often sees her thumbing idly through the books she is tasked to inventory. But what is perhaps surprising is that the company seems to abide these moments of idleness. Unlike the kind of workplace you see in Charlie Chaplin's film *Modern Times* (1936) or *I Love Lucy* (1952), where an assembly line speeds up so much that it swallows Chaplin's body whole or forces Lucille Ball to wolf down the candy she's supposed to be wrapping so that her manager doesn't notice anything unwrapped, the Leipzig warehouse is full of inefficiencies where workers are left idle. Sometimes there are simply no deliveries to unload, for example, so rather than waiting around, the warehouse workers are supposed to sweep inside the management's office. But because the desk workers feel bad about turning them into janitors, they prefer that warehouse workers play hooky instead. The Leipzig management even systematizes slowness, sorting its receiving workers by pace, into a "turbo" shift and a slower, "leftover" shift, the one Geissler ends up in. In essence, Geissler is not important enough to the company to need to go fast—or, more precisely, the "fast" of 24/7 capitalism is defined as a quality of white-collar work (as in the title of the business magazine for entrepreneurs, *Fast Company*). As a temporary worker, she will be gone, anyway, in a few months, and there are thousands of other applicants happy to take her slot.

Digital capitalism contains overlapping temporalities. Fastness overlaps with slowness, often in the same place; how time is experienced is distributed unevenly throughout the supply chain.[5] Lethargy arises from these disjunctions of time. There is, of course, the constant pressure to produce, but how this is experienced—whether as anticipated demand that is planned for years in advance, as seasonal work that lasts a few months, or as inertia and indecision—depends on how wealthy you are. As Sarah Sharma points out, descriptions of the so-called slow movements meant to counteract this fast-paced tempo are always missing the workers who must rush in and out of sight to make slowness possible for others; the wealthy can slow down in a way that the poor cannot. We need vocabulary outside of "slow" and "fast" to describe the contemporary moment, and the temporality explored by Geissler's book manifests itself not so much in the slowness of movement per se, but in an impasse over action. If modern life's accelerations manifest themselves as a depressive weight in which, as political philosopher Wendy Brown puts it, "you cannot move because of the bleakness but you cannot rest because of the anxiety,"[6] this chapter starts with the proposal that digital lethargy is built out of this temporal architecture, where you cannot act but you also cannot be still.

Geissler's book responds to this impasse by exploring a fantasy of achieving some sort of meaningful political action against Amazon. The narrator fills out the story of static and unchanging work by imagining all the ways she could sabotage the warehouse system: by delaying shipments, by working slower, by screwing up each item's inventory category, by damaging products. These dreams of resistance are underscored by the narrator saying: "We're not leaving this book before you've taken action"—ostensibly to herself, but also to the reader, who is included in the "we" and hailed by this call to action.[7] In this way, Geissler's claim is really about how we tell stories about digital capitalism. Stuck inside the book, the reader is set up to expect some sort of precipitating event, some sort of change (the word for plot, *die Handlung*, comes from the same verb for "take action" in German), that interrupts the cycle of the narrator repeatedly doing the same thing. And we in turn expect stories of political change to also involve abrupt disturbances or breaks in the status quo.

To the extent that the narrator's and reader's desire for subversion is fulfilled, it is in a way that is thoroughly deflating. One day, after a season of lifting and moving boxes in a freezing corner of the warehouse, Geissler's narrator begins to mis-scan inventory, slows down as if she is working to rule, and essentially flouts all the norms of job performance. She then skips out on the last two weeks of her contract. Yet what looks like defiance stems simply from fatigue: she is too tired to do anything else. Even worse, Amazon greets these moments of failure with humiliating indifference: an employee calls to offer feedback—"We were very satisfied with your work. Everything was more or less fine"—and then, a few months after the season, Amazon writes her a form letter, sending her another expression of thanks and an invitation for her to apply again for the next contract.

This is not to say that all acts of resistance are pointless. Walkouts against Amazon sparked a successful union drive in New York and won a $15/hour minimum wage in the United States, even as the company refuses to budge from its model of hiring a largely temporary and disposable workforce. But Geissler's pessimism usefully points to a sense that Amazon's opaque management structure makes some forms of resistance almost old-fashioned, to the point that they act less against the corporation than against fellow workers. She frequently observes her coworkers sniping at each other and admonishing each other not to lower their shift's averages: "people don't dare annoy their teachers; they make do with their peers . . . They all want a clear and simple target that's easier to hit than a whole corporation."[8] Indeed, a few years after her temporary job ended, the author returned to the Leipzig warehouse to visit a picket line, greeting some of her former coworkers, while observing that others were wearing "I Love Amazon" T-shirts as they clocked into the factory. Exhaustion, she argues, sets workers against each other—clear and simple targets—rather than against their employer. Seeing them divided between solidarity to one another and loyalty to the company, she admits she often pushed her way past her coworkers in a rush to leave at the end of the day.

If anything, a significant portion of rule-breaking that Amazon employees described to journalist Sam Adler-Bell helped their employer, for instance, by evading safety regulations in order to push one's body harder or by taking cannabis oil to numb anxiety and deaden the crushing workload.[9] The employer's tentacular involvement with its employees' lives is bracingly portrayed in *Seasonal Associate*, as the narrator refuses a position of moral purity. Quoting a writer who admits that "I, too, buy my books from Amazon.com," she self-mockingly imagines demanding a better chair at work, since "I . . . am also a customer of this company."[10] In a bigger humiliation, she calls upon her detailed knowledge of Amazon inventory to one-up a bourgeois couple who flaunt their wealth: she insists she will only eat Marconi chocolates at 30 euros a box ("so worth it")—chocolates that she has presumably only seen and inventoried.[11] The narrator's boyfriend describes her as the only "neoliberal left-winger" he knows.

Amazon operates at such a scale, and with such opacity, that actions that would in another context be disruptive are simply folded back into the system, not just to be tolerated but even to be monetized. Workers disappear from their shifts and take personal time off, intentionally or not, and it all gets tracked for feedback. These "disruptions" mostly seem to be part of the system's design or, in the case of the letter inviting Geissler to rejoin Amazon, a signal that the view from the HR computer system is radically different from the view below. Posters on an online forum for Amazon Fulfillment Center workers, for example, describe how Amazon's byzantine "voluntary time off" system is actually a way for Amazon to hire more associates than it needs, and then push or coerce tired workers into taking time off when there's no work for them to do.[12] This way, Amazon can not only claim government subsidies for having employed more workers, but gets a highly scalable pool of manpower for surges and lulls in business. These strategies reinforce the fact that Amazon is not a traditional retailer. It is also the largest provider of cloud services in the world, renting capacity to companies such as Netflix and Snap to absorb sudden spikes in Internet traffic. In that domain of its business and more broadly, it is at heart a logistical fulfillment system ("logistics as a service") for other sellers besides itself. It rents out not just warehouse space and package delivery but—importantly—a labor force that can scale up or scale down when demand intensifies or slackens with the season or from day to day.

Further, this system's opacity means individuals caught within it can lose sense of cause and effect. Amazon is an example of a wider managerial (and even societal) shift from causal or rule-based mechanisms—for example, "You're fired if you're X minutes late Y times"—to the correlative mechanisms of big data—"Please come back for another contract because the computer says you fit its pattern of productive workers."[13] One acts, and perhaps nothing happens, or perhaps something has happened. But if there is no sense that what one does in the present has any implication for the future, then the idea of the future that underpins much of contemporary life—a future that is saved for, or planned for—must be altered, even suspended. For the "temporary" workers of Amazon, many of whom work one temporary gig after another, the future simply repeats and iterates on the present. What the reader of *Seasonal Associate* is left with is simply the body wearing down, of time taking its toll.

Historian E. P. Thompson famously showed that industrial capitalism changed the way that people thought about time: "Time is now currency: it is not passed but spent."[14] We see this logic in Amazon's exhortations to its temp workers to put in the time to jockey for a permanent position, where one can then move up in the ranks not just from an hourly position into a salaried position, such as that of a manager, but, theoretically, from blue-collar labor into white-collar information technology work. ("We at

Amazon think every day is a first day," recites the American MBA student who conducts Geissler's narrator's training—language from start-up culture that still opens Amazon's job ads.[15]) But the form of Geissler's book allows us to imagine or remember what happens when Thompson's formulation is reversed, and time is no longer a resource to be marshaled, invested, or spent, but simply something to be passed.

This is evident in the time the narrator uses to detail the plodding routines of daily existence, most of which are still tied to her job: how to apply for welfare, how to get a sick note from the doctor, how to commute, how to have a "sensible employee's sleep" before working the next day. "Incidentally, you've now started talking about the weather," she says to herself, because this most banal of subjects is what she becomes most attentive to as the winter cold begins to permeate the unheated corner of her warehouse.[16] To write about ordinariness, Geissler has looked to influences such as Chantal Akerman's short film *Saute ma ville* (1968), which parodies the domestic regimen of a housewife in her kitchen with manic exaggeration (smearing shoe polish on her legs, for example), and the British social research project Mass Observation, which sent questionnaires to its respondents from 1937 to 1949 on a variety of topics in their daily lives that were simultaneously intimate and public, from grooming habits to the "Lambeth walk" dance craze.[17] Both projects redirect the focus from the macro to the quotidian and suggest that the motions of domestic labor are extensions of the motions of work. In turn, Geissler's inability to do domestic labor—at one point, she admits choosing to sleep over wanting to attend to her children—speaks to the way that work's exhaustion breaks down these routines, even as there is nothing *but* routine to turn to when passing the time. If spare time is, in the words of Humphrey Jennings, one of Mass Observation's founders, "a time when we have a chance to do what we like, a chance to be most ourselves," fatigue mutes Geissler's ability to imagine what she likes, even her ability to be a self.[18]

This is why the real subject of *Seasonal Associate* is less Amazon than fatigue. Fatigue is isolating, and it deflates not only the individual's relationship to family, but also the efficacy of collective action. Quoting the philosopher Byung-Chul Han, she writes: "There's no way to form a revolutionary mass out of exhausted, depressed, isolated individuals."[19] At one point, the narrator imagines a crowd of warehouse workers walking out, but she quickly understands that they are too exhausted to have learned any new skills in their spare time, and barely able to organize, let alone strike. Even if they did leave, she thinks, "they'd be nothing but the cause of a minor temporary personnel shortage, expressed in a single sentence on the company website: Due to poor weather conditions, we currently anticipate delivery delays of up to three days. We apologize for any inconvenience."[20]

This pessimism is what sets *Seasonal Associate* apart from the muck-raking books that critics typically group it with, such as Upton Sinclair's novel *The Jungle*.[21] While Geissler may appear to provide a how-to guide on how to expose exploitation or how to act, a closer look invites us to ask, instead: What is action inside the structure of exhaustion? By the end of the book, the narrator is not sure that any action, broadly defined, has happened: "I wrote: We're not leaving this book until you've taken action. / I'm not sure. Have you taken action or not. / Yes, you have. / We'll see. / Let's stay in touch."[22] The book ends without resolution. And this twinned state of temporariness and temporal out-of-jointness also asks us to rethink ideas that are normally bound to a progressive sense of time, such as agency. As anthropologist Katherine Stewart observes, writing about the texture of ordinariness, "[a]gency can be strange, twisted . . . passive or exhausted. Not the way we like to think about it. Not usually a simple projection toward a future."[23]

What might a "passive," "exhausted," or—to use the term this book proposes—"lethargic" capacity look like? Perhaps the best clue is exemplified in the quote above, when Geissler splits the narrative into a dialogue between "I" and "you": "I," the present-day narrator who is in dialogue with the "you," the former worker at Amazon. (The difference is even more magnified in the original German, where "you" is designated by the formal *Sie*, a distanced and respectful mode of address pointedly avoided by Amazon Fulfillment Germany, which instead has adopted its American parent's culture of first names and feigned friendship.) These two voices diverge in the book as much as coincide: at one point the "I" goes to have wine with friends in the present, "abandoning" the "you" (and the trouble of writing about Amazon) to her work. Fatigue, the philosopher Emmanuel Levinas has argued, means that the body has become a being "no longer in step with itself"[24]—and in the book, the "I" and the "you" are often literally out of step: during one commute home, the "I" yammers away about change at an exhausted "you," who ignores her. "Perhaps I chose a bad moment. I walk a little slower, letting you walk ahead."[25]

Fatigue's out-of-syncness leaves its mark on the memoir's form. Rather than simply recounting what happened, her present-day self is also talking, too, frequently without knowing what's happened to her past self. This talking-past, out-of-joint quality means that the book unfolds as a process of waiting for the self to catch up to itself. "We'll skip a few more days, as we've been doing all along."[26] Fatigue undoes both the physical boundaries of the "I" and also the temporal training that constitutes the individual: the self that we construct so that we can tell others; the self sequenced on a resume or a timeline or a feed that consists of one image or accomplishment and then the next; even the social-ized self as a coherent subject that retains authenticity over time. (This self-forgetting is at times literal: the book, which was reconstructed from sticky notes that Geissler jotted at the warehouse, works with the faults and gaps in her own memory.) In an interview about her writing process, Geissler recounts that her initial draft of the book was a simple recount-ing of events in first person. But simply inhabiting the internal thoughts of the narrator as warehouse worker was claustrophobic, and she found that switching from first person to second person and then adding an occasional first-person voice allowed the book to better reflect her lived experience.[27] The "you" forgets itself, and becomes an "I," for a spell, allowing one to disassociate, drift. This cleaving between the "you" and the "I" voices generates much of the book's internal rhythms. In a book otherwise devoid of surprise or notable events—each day in a warehouse shift is much like the previous day—incidents of lethargic forgetting become the way that the narrator registers and mediates time.

Fatigue may seem like a state of inaction: a period of being too tired to do anything, a period of waiting for one's energy to be restored. Instead, it makes space for other ways of inhabiting one's body. Comments cinema scholar Elena Gorfinkel: "Tiredness is not inaction but instead is a reflexive holding in abeyance, the body waiting for itself to recharge, reenergize, or waiting for a shifting desire, drive, event, or an approaching relation to the world. This event may never arrive. But it is the qualitative expectancy of waiting that infuses the banality of tiredness with its potentiality."[28] In other words, tiredness is itself a type of potential. After all, weariness is not always opposed to action or agency; it can work alongside action to produce the Silicon Valley mythos of scientists pulling all-nighters to deliver a revolutionary new product, or, alternately, the lethargic loosening and forgetting of temporal and social binds. Tiredness can drag the body back to the present, rather than allowing it to race ahead to the future; it focuses attention on the process of waiting, and the inchoate energies that adhere within.

Gorfinkel invokes the theories of the modernist filmmaker and writer Jean Epstein, who argued that fatigue is not the depletion of a body (like a drained battery) but rather a new symptom of modernity: resulting from an excess of stimulation, fatigue signifies "inchoate potential" or "unactualized potential."[29] It's a fascinating reversal of our usual associations with fatigue: it's something like the world-weary phrase "I could write a book," which, as Stewart explains, means that the speaker won't ever write a book and doesn't know how to, given the sheer amount of experience they have been storing up. The literal expression of those experiences—the actual book, as it were—isn't the point of the phrase; the point is the overflowing affective charge (albeit thwarted) implied within. "Wouldn't know where to start and how to stop."[30]

At least Geissler is lucky enough to have worked a single seasonal contract with a start and an end date, which lets her book have a start and an end point; we couldn't say as much for most of her coworkers. Were we to sample their stories, though, they might show that weariness is a heightened, if strange, form of attention and noticing, the attention shown by an insomniac who is unable to filter or shut off perception and thought. "Wouldn't know where to start and how to stop" would describe a state and a kind of writing that would be realizable only through exhaustion, rather than through will and wakefulness. Lethargy isn't absence—just the opposite, in fact: it's both feeling overwhelmed and being overwhelmed with the problem of feeling.

BECOMING INVENTORY

As Geissler waits for her contract to terminate, she writes: "You close your eyes and sit in an ocean of time. . . . Your internal screenwriter demands a big closing scene, but none comes and it never will. No grand gesture, nothing that might be good for a minor showdown."[31] Her employment contract is so open-ended that even the official "end" of the employment will pass by without closure or even a response from Amazon. What happens, instead, is a feeling of having more time than one knows what to do with, of time accumulating into an ocean. Geissler's exhaustion acts like a microscope, revealing that waiting isn't an interim station on the path elsewhere but an abundant world teeming with life. She begins to see herself more and more like the people around her: a man who rides the tram all day long, a "professional passenger"; a street musician who sings the wrong lyrics; a beautiful old woman who fishes bottles from trash cans; a neighbor who cleans houses for free because that's what the welfare office wants.

This is in sharp contrast to how a corporation sees time. The supposed crime of stealing time from an employer was formally codified in a 1983 time-use survey, writes sociologist Laureen Snider, spurring outraged managers to decry the "time bandits" and "time thieves" who were engaging in "America's Biggest Crime: . . . idle chatter, hours spent on the phone with family and friends . . . reading on company time . . . over-associating with co-workers."[32] (Amazon, its associates report, is notorious for using time theft to justify terminating contracts.) Geissler's response to this supposed scarcity of corporate time—and, implicitly, modernity's process of making time something measurable and commodifiable in the first place—is to depict time within Amazon in all its abundance. For Geissler, time is neither spent nor saved (the implicit ideas that undergird a 24/7, always-on capitalism, whether boosterish or critical) but is instead passed.

To wait for an "ocean of time" to drain away, Geissler begins listing things in the warehouse: horse-themed calendars, mugs with a portrait of George Clooney, table lamps, green rabbits named Mombel Wombels. To wait, in other words, is to watch. Geissler's work sorting inventory, where she sees the dislocation between the best sellers and the deadstock that accumulates and then sits untouched, such as a glass bathtub duck or a hair dryer named Alpine Hairdryer Grossglockner, produces a drama of stagnation. At one point, she comes across an entire pallet of nearly identical books titled "The Marketing Secret for Dentists," "The Marketing Secret for Catering Companies," "The Marketing Secret for Fitness Studios," "The Marketing Secret for Real Estate Agents," "The Marketing Secret for Bakers," "The Marketing Secret for Roof Tilers," "The Marketing Secret for Metal Workers," and so on. These books were doubtlessly written in anticipation of Amazon's search engines (and its segmentation of consumers by marketing categories), even as virtually nobody will read them. Geissler's list hints at the disposability of those books, of the excess that must be written to hit its target, like a spam email that is opened by only a tiny fraction of its recipients. Amazon, the self-styled "Everything Store," first made its mark by stocking items on the end of what businesspeople call the "long tail"—stocking things people rarely buy or want—in contrast to physical bookstores that primarily kept top sellers in stock. Her view from the warehouse floor shows how the ebb and flow of consumer shopping and consumer trend is itself a form of passing time. "This is where the stock comes to rest; the stock really seems to be sleeping and not actually for sale. . . . You see a dust-coated stock museum; you like it. The things on the shelves, silence reigning around them because no one is here to collect them and send them to the customer, radiate sobriety. . . . The products look like retired former workers for this global corporation."[33]

During her work, Geissler develops a curious solidarity for the objects that she handles. At one point she compares her own travels to those of a carton of commemorative porcelain mugs, "made, printed, and packaged in China; then offered for sale in France; and now shipped from Amazon France to Amazon Germany as seasonal specials."[34] The mugs, she realizes, have traveled further than she ever has. But both inventory and worker occupy seasonal and transient parts of the supply chain; one is a seasonal special and the other is a seasonal associate. Though Geissler only occasionally touches on the complex logistics of "made-in-China" sourcing, we can nevertheless understand the distribution of value on different parts of the supply chain, where manufacturing labor is extracted from certain states (here, China) and sorting and distribution from others (in her case, the deindustrialized states of former East Germany). The mugs are congealed human labor, and it is through their damage that they come to have a sort of animacy: "sometimes you drop a product, and then you get a shock and hope it's not broken . . . it's a matter of treating the products roughly, so that they don't rise up above the workers. You'll learn that soon enough, but you'll never be much good at it."[35] The disheartening realization here is that the worker is barely differentiated from the products that move through the system. Indeed, as philosopher Achille Mbembe claims, as a result of capitalism's animist worship of commodity objects, "many people want to become objects, or be treated as such, if only because becoming an object one might end up being treated better than as a human."[36]

Just as Geissler likens product deadstock on the shelves to "retired former workers," she in turn compares the labor force to inventory that the company stockpiles, manages, tags, and controls. A strike is tantamount to "delivery delays" and "bad weather"; the state job agency fills Amazon with temporary agents who live two or three hours away from the warehouse, a logic that makes sense only in the curious geography of supply-chain capitalism, where parts and packages are frequently sent five states over to be re-sorted and then sent back. The manager who tracks workers trying to clock out early acts as another form of "inventory control" for Amazon's human resources.

To become inventory is to experience waiting as a state where "doing" is reversed into "something is about to happen to me." This is not passivity but lethargy. The subject slackens from a doer into an object of forces out of her reach: algorithms that calculate employer performance or production quotas; algorithms that respond to consumer desires and wants by sending ramshackle boxes of odds and ends into the warehouse and into her hands. Controlling none of it, she waits for it to happen. In addition to the physical effects of fatigue, this waiting is another avenue by which the lethargic body becomes an object set adrift from her role as active agent, even from herself as subject. Rather than becoming the agent of a "minor showdown," the lethargic body watches things happen to itself as an object—the slow wear of hands repeatedly frozen and thawed, a back hunched over: fatigue, in the sense of material fatigue.

Geissler imagines herself and her coworkers replaced by robots in a few years, but her story reveals more than predictions about growing automation or the familiar observation that Amazon, or any other company, objectifies its workers, treating them like disposable machine parts. The image is part of her broader exercise of returning to, and re-experiencing, her own objecthood. The scholar Anne Anlin Cheng has asked: "How do we take seriously the life of a subject who lives as an object?" The challenge of her question is to take the subject on its own terms, rather than to cure it of its objecthood. Cheng argues that personhood is often suspended between thingness and humanness; indeed, "this complicated congealment may be what is possible in a life of precarity."[37] Deriving her analysis from the way that Asian women have often been fused with or entangled with ornament, commodity, and artifice, she offers an example of an American journalist who pejoratively describes Chinese women gymnasts as fragile as china (porcelain); Cheng shows that the historical precarity of these women has forced them to endure *through* their thingness, for example as spectacles to be looked at. But the same description of "person thingness" applies to other hybrids: artificial intelligence, in her example, or the relationship between one's data body and one's biological body (considered in the next chapter) or simply, in this case, the disposability of workers who serve as human infrastructures for digital capitalism.

Though Geissler is not Asian, her reanimation of the porcelain mugs produces a fleeting moment of intimacy between the two continents, a moment of contact where she realizes her disposability and the commodity comes, briefly, alive.[38] For the supply chain is also a hybrid between animated things and deanimated persons, and traveling its links allows Geissler to inhabit her thingness: her mug-like disposability, her lethargy, her ability to be purchased, her embeddedness inside the supply chain, as well as her sense that things seem to have a pathos of their own. But this realization is not a matter of Marxist alienation; it is an ethical stance. Let me explain: the division between subjects and objects is, as scholars of race point out, the very goal of racialization, which makes some bodies objects and makes other subjects seem universal. By objectifying people of color, whiteness shores up its sense of subjectivity, allowing it to be, as performance studies scholar Lara Shalson puts it, the imagination of a self as the "one who acts."[39] Yet the idea that one can purify a body of its thingness is the impossible fantasy of white supremacy. To instead inhabit one's objecthood, as Geissler does, is also to unsettle the ground on which white supremacy is built. Even though humans are both object and subject at the same time, we continue to see universalizing models of subjectivity permeate civil society, which often imagines a dialogue between subjects in the public sphere.[40] For a person of color or an immigrant, however, speech is framed as surprising and exceptional: they are either condescendingly described as remarkably "articulate" or, conversely, dismissed for not speaking in a "civil" fashion. Indeed, enslaved persons historically had little access to the category of speech; their owners complained that they produced only incomprehensible noise.[41] The public is a public for subjects, not objects; it excludes persons who are designated in advance as objects.

Rather than repeat this mistake, Shalson proposes an alternate model of relation that leans into a body's objecthood. Endurance, that is, "the capacity of bodies to be acted upon," points to something rarely visible if we consider only a body's publicity.[42] Consider the 1960 lunch counter sit-ins in Greensboro, North Carolina, which began when four Black students who were denied service at a segregated Woolworth's refused to leave. Viewing the press photos of the sit-ins, one might anticipate the confrontation: the way that violence is about to descend on the student's bodies, the social context of the moment, the way the bodies are protesting discrimination. Yet Shalson describes the scene differently: the sit-in protesters "exerted an agency separated from, rather than wed to, violence."[43] And, indeed, there are other things besides violence and protest to notice. Kevin Quashie, a scholar who has written on how Black subjects are overidentified with expressivity and resistance, points out that these students "are not (only) in a stand-off but they are also reading; their minds are likely wandering over many things, including the words on the page . . . one should not disregard the intimacy of their posture . . . the specific loveliness of these two young men."[44] Endurance here is not simply a means to reverse the received framing of violence but an exploration of bodies that receive action.[45] That separateness, even opacity, of intent is its own capacity. Put another way, objecthood is best explored in its own terms; even as one kind of public life is curtailed and framed by violence, another kind of life is lived, laterally, within those constraints.

Where lethargy is concerned, endurance can serve as its own field of potential, separate from protest or resistance (even if and when they coincide). For the lethargic process of waiting in digital capitalism is not a process of waiting *for* a (political) future to arrive but a process of endurance, of waiting or remaining *within* an unbearable situation.[46] If we typically think of agency as located in the subject of the sentence—A acts on B—lethargy is, instead, the capacity to receive action; it resides in the object.

The question of who is acting on what reaches beyond grammar and into one of the central debates about the technology industry. It is an increasingly common observation that we are used *by* digital platforms, rather than users *of* them; critics use this formulation as a way of decrying our vulnerability to those platforms. But to simply reverse that formulation only restores and reasserts the sovereign power of users to issue commands to servers, rather than accepting their own objecthood. Indeed, the binary of subject and object is reinscribed and exacerbated by the ways that digital platforms connect and transform populations into active "users" and passive "services" or "servers" who wait to be activated by their users. These platforms free wealthier programmers and consumers from drudgework and give them a feeling of agency and creativity, even as they limit communication with their servers to protocol. In doing so, they can suppress not just the server's identity, but their ontological status as human or AI. (In 2016, reports on Facebook's "trending top stories" feature showed how confusing the digital supply chain can be, even to media scholars, who assumed that Facebook used algorithms to mimic humans and generate its news feeds, until a leak revealed that they were actually humans employed to act like algorithms acting like humans.[47] As of the time of this writing, the news feed feature has been discontinued, and those humans fired.)

Today's digital supply chain relies heavily on logistical technologies that coordinate bodies alongside their algorithms or warehouse robots. Even as the fulfillment center workers, in the words of writer Alec Recinos, "turn familiar objects (books, mugs, toys) into abstract parcels . . . in order to facilitate their circulation,"[48] digital algorithms similarly turn human workers—pickers, microworkers, content moderators—into abstract objects that circulate. In computer science lingo, an object is a way of containing and thus abstracting internal data and code, and so platforms for hiring work on demand often literally turn a global workforce into objects that other programs can use—a process that Amazon's founder describes as "human[s] as a service." The true objects of logistics are human— and those are disproportionately persons of color. As of 2020, 71 percent of Amazon's front-line warehouse and call center workers, for example, were nonwhite.[49] Workers are inventoried and tracked and even efficiently packed into the aisles of Amazon warehouses in a historical echo of how logistics was, as theorists Stefano Harney and Fred Moten write, "founded with the first great movement of commodities, the ones that could speak. It was founded in the Atlantic slave trade, founded against the Atlantic slave."[50] The efficient containerization of an enslaved person's body into the hold of a ship was the first innovation in this dire trade.[51]

While the violence of the Atlantic slave trade and the exploitation in the digital economy are incommensurable—glibly collapsing the two cheapens discussion of both—they are both products of the racial capitalism that, for centuries, has endeavored to strip humans of subjectivity and turn them into services. For the "innovations" of racial capitalism extend far beyond the technology of enslavement; they have also included the logistical techniques of abstraction, surveillance, biometrics, fungibility (e.g., containerization), and exchange.[52] The writer and philosopher Sylvia Wynter has described the "reduction [of Antillean bodies] to the status of *pieza* (i.e., of being so many units of extractable labor capacity)," a term that referred to the standard measure of a male slave of about twenty-five years age ("Two young boys, for example, would then make up one *pieza*").[53] Racialized bodies continue to be raw materials for extraction: for example, in 2018, Guangzhou-based CloudWalk Technologies signed an agreement with Zimbabwe to mine a national ID database for training Chinese facial recognition systems on Black faces.[54] But we also now see labor measured in terms of Human Intelligence Tasks (HITs), another imprecise yet seemingly objective unit of making human labor fungible on platforms such as Amazon Mechanical Turk. The systems that extract labor for digital capitalism are tailored to each locale: while digital microwork may take place in India or Bangladesh or Kenya, elsewhere on the supply chain, it may be the manufacturing of electronic components by Asian women, or customer service work in Mexico. Yet the very divisions of a global labor force into tasks informed by racial hierarchies, such as seeing the "nimble fingers" of Asian women as "naturally" suited for assembling electronics, or promoting microwork as a way of "saving" impoverished Kenyans and Ugandans and thus equating it to a form of benevolent aid, show how a logic of coloniality continues to govern digital populations today.[55]

Even as digital platforms split populations into users and servers, wealthy consumers in the Global North and poorer servers in the Global South are two different parts of the digital supply chain. The irony is that all of their data bodies are products, captured and delivered to finance capital, whether through the marketing of their selfhood or through the extraction of their "human intelligence." Users and laborers are all ensnared in the same net, suffering different and disproportionate, but still lethargic, burdens. The potential of lethargy—and of inhabiting one's objecthood—is to remind us of the intimate relationship between users and servers, however separate they may seem. For, ultimately, solidarity within the undercommons is possible only when one relinquishes an attachment to the idea of a person as a doer and agentive subject, and endeavors to understand the perspective of a commodity that can speak but is not listened to. Lethargy, paradoxically, can be a choice; one can surrender one's attachment to agency and accept the object's vulnerability: "He is the shipped. We are the shipped, if we choose to be" (Harney and Moten).[56]

I have started *Digital Lethargy* by considering a work of literature for a reason: because Geissler's memoir, which places us not just in the shoes of a temporary employee but in the position of inhabiting the temporariness or ongoingness of lethargy, offers a useful way to think about the *form* of lethargy. It attempts to fit a situation that is endless into a narrative that has a beginning and an end. Geissler uses the start and stop dates of her one season on the job to frame her insights into time: the time of a (human or nonhuman) object's transit within just-in-time logistical circuits, the feeling of temporariness and stockpile. This helps us ask: How is "the life of a subject who lives as an object" located in the time of narrative? How do we explore inventory and stockpiles and seasonality, as opposed to the growth and development of characters? What is the narrative of an object that does not visibly act but simply waits to receive action? How do we tell a story about standstills and impasses, in short, about lethargy? If we could do this, we might offer a very different history of the present—told not only from the perspective of or about its subjects but also from the perspective of its objects.

NO FUTURE, FOR NOW

When I was in my twenties, I often enjoyed watching films that were boring—films where very little happened, and where character development was not a priority. In one, I remember being lulled half to sleep by the gentle background din of plates inside a rotating restaurant; in another, the listless dread of three family members squabbling as they waited for the end of the world. It was not until much later that I realized what those films spoke to in my own life: a feeling not of boredom but rather of stuckness. In a collection of personal essays titled *Minor Feelings*, the poet Cathy Park Hong defines her book's title by describing a body of literature that comes out of a lack of change—"in particular, [the lack of] structural racial and economic change . . . Rather than the individual's growth, the literature of minor feelings explores the trauma of a racist capitalist system that keeps the individual in place."[57] Even if minor feelings vary from person to person or moment to moment—paranoia, hostility, shame, melancholy, ungratefulness—they share a common problem: involuted, they gnaw away like chronic illnesses that one is only partially aware of, and thus take a while to manifest, if they come out at all. Indeed, one of the reasons that they are involuted is part of the feedback loop of dissonance that gaslights the bearer of these feelings and thus generates more minor feelings. This dissonance was something I increasingly recognized in my life, too, particularly around the entrenched American attribute of optimism, where, Hong recounts, "you are told 'Things are so much better,' when you think, 'Things are the same.'"[58]

This optimism has been baked into the ethos of Silicon Valley, where I once worked as a network engineer, and which rarely tires of telling its consumers that "things are so much better." Its evangelists claim that a knowledge economy can make the world a better place, typically by liberating us from older and often repressive power structures—a conceit that cultural historians have dubbed "techno-optimism" or the "California ideology." But even though these grandiose claims have largely soured for the public after revelations about dataveillance, monopolization, and worker exploitation, the industry continues to use optimism as a business strategy to reach individual users, to make the feeling of a future that is just about to arrive an integral and desirable part of how we experience life online. We look forward, both consciously and unconsciously, to the next update in technology that will solve our problems as much as the next video on our feed, to the next recommendation and thus to the next iteration of the algorithmic "self" that is returned to us, to the next moment of contact from a friend or colleague, which arrives in the next notification. This anticipation produces a sense that everything is constantly changing, and this constant flux is what feels like and what we now call "liveness."

As digital media scholar Neta Alexander argues, the open secret of techno-optimism is that it is really a tactic of making users wait for things; indeed, waiting is the "underlying logic, ideology, and business strategy" of digital capitalism and of neoliberalism more generally.[59] A host of tech companies have co-opted our discomfort with waiting (even as they produce this discomfort), for instance, by offering work and educational smartphone apps that make downtime more productive or by helping make that downtime more social or more connected.[60] Even when interruptions or failures on digital platforms occur, such as a video stream stalled in mid-load or disappointment with a company who has mishandled our data, our attachment to what is always about to come—even toward the wait itself—strengthens nonetheless.[61] For example, today's critics often respond to Silicon Valley's failures by placing their hopes in the next generation of technology, such as a more decentralized network or additional encryption. This faith closely resembles an affective structure that cultural theorist Lauren Berlant calls "cruel optimism," where our attachment to a "good life" promised by neoliberalism is in fact the obstacle that prevents us from attaining it.[62]

Because digital capitalism is typically seen as a progression or intensification of neoliberalism, Alexander and others study it, following Berlant's model, through emotional states that refract that attachment to the future, such as anxiety. Yet a diagnosis of (cruel) optimism is a misdiagnosis for many. It leaves out subjects who were born too late to experience an expansionary economy, those who have never been overly attached to the "good life" promised by (digital) capitalism in the first place, or subjects for whom narratives of change feel dissonant or unconvincing. Add to this list, too, populations that have been re-engineered by big data, for whom the future has already been pre-filtered and constrained, if not entirely curtailed, by algorithms. To offer just one example, algorithms currently inform parole decisions and where police are deployed, such that these decisions can be made "objectively." Yet because the algorithms depend on extrapolations from historical data, they replicate the racial bias that previous decisions reflect, such as incorrectly assigning Black defendants a higher risk factor for parole, and over-policing Black neighborhoods.[63] Prison abolitionist Jackie Wang points out that technology promising to predict crime isn't just inaccurate; it also enacts racial stratifications,[64] ensuring that someone who is Black has a higher chance of going to jail due to decisions made well before they were even born. On a more mundane level, automated credit decisions and search engines target results based on one's imputed demographic category, and accordingly deliver only the opportunities perceived to be relevant to that category. Because these processes create futures that are arguably reiterations of the past, these predictive loops are remarkably static, perpetuating economic or physical immobility. Before visitors even begin browsing a website, an analytics company such as QuantCast uses attributes such as geographic location and prior web history to racialize them.[65] These populations have arrived after choices have been decided for them; they have arrived too late on the scene to make a difference; they are often, to invoke Mbembe, subjects for whom "the end of the world has already happened."[66]

To arrive "too late" is to lose a future that one has never known or has never had access to. (Here digital lethargy takes on a family resemblance to its historical cousin, melancholia, where the sense of being unable to define what exactly has been lost makes the future an "unknown loss."[67] Perhaps these minor feelings accumulate as exhaustion or perhaps they go unnoticed; either way, they mark the subject as lethargic while being nameless themselves.) Thus when critiques aim to intervene and correct a logic or a crisis brought on by contemporary technologies, they rarely consider the temporal framework with which we ask these questions. Consider the common complaint that new digital technologies, such as automated facial recognition, have dragged us into a regime of mass surveillance. In response, sociologist Simone Browne argues not only that surveillance is "nothing new to black folks," but that we can only understand how surveillance operates *through* its ongoingness: "to see it as ongoing is to insist that we factor in how racism and antiblackness undergird and sustain the intersecting surveillances of our present order."[68] A better understanding of the contemporary moment requires setting aside the assumption that technology has brought about radical societal change (however good or bad). Instead, we should better account for stasis, for the fact that for many, things are the same.

Racialized subjects inhabit alternate temporalities in relation to technology.[69] On my part, I've always had the privilege of seeming to know what I am talking about when I talk about computers and programming code, even though I haven't written a line of code in over a decade—most commonly because people think that, to quote a character from the TV show *Futurama*, "you're from one of those ethnicities that knows about technology."[70] But I also remember a strange week early in my career as a professor, when I got accosted at a poetry reading by an audience member, who wanted to know where I was really from (the "correct" answer, at long last: China), and as a result whether I ate dogs, and later, by a senior colleague in the hallway of my university, asking whether I could help fix their new computer. Was I primitive or was I from the future? I learned that this strange temporal contradiction is a result of what Greta Niu, Betsy Huang, and David Roh describe as "techno-Orientalism," a way of positioning Asians as outside time in order to reaffirm the West's centrality.[71]

To make sense of these contradictions, and to better describe populations who are "after the end of the world," I found myself turning to the lyric that the Sun Ra Arkestra sings to open the 1974 film *Space Is the*

Place, a lyric that cinema scholar Kara Keeling uses to describe this muddle: "It's after the end of the world. Don't you know that yet?" For Keeling, such works of Afrofuturism may initially seem to make a claim to a different future or another world, however speculative (or even cheesy): in the film, Sun Ra has founded a settlement for Black people on a psychedelic planet, complete with Lorax-like trees and floating life-forms. But in the wake of the unthinkable trauma of the Middle Passage, Keeling writes, Afrofuturist "investments are not in the future, or even *a* future, but Now—a now that Afrofuturism constantly destroys through its insistent discordance with it."[72] That destruction is a way of clearing space for whatever might come next, but that clearing must happen continuously, as a way of registering discord with it. For all the film's spectacle, one quotidian exchange captures this discord best: Sun Ra and a drunk man converse over their shared status of being nothing to society; the drunk man tells him that he has been doing nothing for "quite some time, quite some time," and in response, Sun Ra proposes to hire him as an expert in "doin' nothing."[73] The offer is a way of both making something out of the temporal stillness of Blackness and disinvesting in a world that produces that temporality. As Sun Ra puts it: "The first thing to do is to consider time to be officially ended."[74]

This insistence on doing work on the now, rather than dwelling in the possibility of what is to come, is in conversation with philosophers of pessimism. Sun Ra's exploration of the position of "nothing" and the time of Blackness is also a key starting point for a group of Black intellectuals such as Frank B. Wilderson III, Jared Sexton, and Hortense Spillers. Loosely referred to as Afro-pessimists, they argue that Black people have always been defined first as objects, and, as a result, Black culture must exist without the endless liberal promises about reform and participation that have in actuality been ways of deferring systematic change for centuries; indeed, it must exist "without a future promising anything different, or, rather, better."[75] Stated simply, Black life must go on without depending on a future promised by others, and on states reliant on such futures, such as hope.[76] Afro-pessimism's insistence on the fixed status of Blackness as slave is not without its share of detractors, who point out that they tend to reduce the social and cultural history of the African diaspora into the African American experience, and also that flattening each and every Black person into an instance of Black death makes for political nihilism.[77] Despite these problems, and despite the singularity of Blackness in Afro-pessimist thought, I find that it offers a set of tools for thinking that apply more generally. I am compelled by the way that Afro-pessimist thinkers wrestle with giving aesthetic form to this flatness (one critic describes Wilderson's own solution as writing a memoir that depicts "life as a series of cutouts"[78]) and who also reframe the politics of repair as something that is largely foreclosed to—or at least unable to account for—populations that have been abandoned.[79] Indeed, repair is often precisely the problem, a way of selling a future to colonized or disposable subjects as something just about to come—if only they will be patient.

Pessimism is a useful counterweight to the optimistic thumb on the scale, particularly when it comes to talking about technology. It shifts our attention away from the future as something that one saves for, prepares for, or otherwise is attached to, and replaces that orientation with an attention to the seemingly "dead time" of an object trapped in the present. And if a pessimist seems unhelpful or unconstructive by questioning the questions that most critics ask or by questioning the critic's implicit optimism ("Things would be so much better, if . . ."), this is its own intervention into the narrative of criticism. "Instead of blaming pessimism, perhaps, we can learn from it. Rather than hiding from the ugliness of the world, perhaps we can discover how best to withstand it" (Joshua Dienstag).[80]

Lethargy, I have been arguing, comes about from attempting to withstand the world: it describes endurance as a way of life, even as a way of focusing intensely on the now. It's this sort of occupying time that feels especially salient to today's environment of underemployment and precarity. The incarcerated do time; the sans-papiers wait for papers; and even the state of waiting for something to happen is professionalized into lower-income, often sub-proletarian work: security guards and watchmen, night-shift caregivers, soldiers told to "hurry up and wait." It also encompasses the act of looking for a job: people on welfare and day laborers spend their days queuing up and waiting for some work to find them. The digital version is the platform employee who spends most of their time waiting for the "ping" on their smartphone that announces a new food delivery or a new ride to pick up.

Even as pessimism suggests an unchanging same, varieties of pessimism have themselves changed, and lethargy aligns with its contemporary forms. Indeed, pessimism is a common motif in protest movements; in the last half-century alone, "No Future," the closing refrain to the Sex Pistols' 1977 punk single "God Save the Queen," became an anthem for a generation of disaffected youth,[81] and queer theorist Lee Edelman's own *No Future* (2004) called for an end to patterns of reproducing (metaphorically and literally) a heteronormative future that is "mere repetition and just as lethal as the past."[82] But the affective and political charge of pessimistic movements today differs from these earlier movements. Sociologist Rebecca Coleman has examined a series of allied movements in Europe, such as the Spanish Juventud sin Futuro (Youth without a Future), that similarly use the idea of a generation lacking a future to protest conditions post-austerity. Yet, as she points out, earlier movements typically assumed a linear or reproductive future that improves on the present but that the dissenters then reject. In contrast, with contemporary movements, not only is there "no better future to reject," but instead the present is delinked from its thrall to the future, becoming political in its own right: "The dynamic flux of pessimism involves 'not just a reaction' . . . [but] the creation of a politics of the present, focusing attention on how the day to day requires change, now. Such a politics of the present sees the present not as a step towards the future, nor as a sphere onto which it might be possible to project backwards in order to assess whether or not a problem has been effectively addressed or solved."[83]

Each of the temporal reorientations I have examined in this chapter—the way that fatigue stretches and undoes how we plan and spend time, the way that waiting is also a state of attentiveness, the way that thwarted action contains its own inchoate potential—is my attempt to map out different aspects of the present's "dynamic flux." By calling these dynamic moments "lethargic," I have understood lethargy as simultaneously "flattening and enlivening" (as Coleman describes pessimism). I have also described the pressure of having, maintaining, or working toward a future as a habit that can be at least temporarily forgotten. Lethargy is a waiting without anything to wait for.

Thinking about the present as a state of lethargy—and thus, as I've tried to argue, to embody a pessimistic dwelling in the now—is as valuable, and arguably as necessary, as crisis-driven solutions and interventions. For populations that exist in a world of "dead time" or have lost faith in the very idea of futurity, or simply for everyone else who "uses" them, this reorientation is particularly important. To examine exhaustion, waiting, and the passage of time is also to realize that a rich and thriving set of capacities adhere to those states. And because lethargy is affective, it is a shared rather than individual condition: it is simultaneously involuted and nameless, social and dynamic. As Harney and Moten remind us, to choose objecthood is also to find the sociality and closeness that occurs between objects, within the hold, "the feeling of a thing that unsettles with others."[84]

Far from a null state that should be redeemed and brought to liveness, lethargy is the forgetting and loosening of the straitjacket that futurity places upon the present. If exhausted, she is dragged backward by her own body; she is forced to wait for herself, but this form of endurance is a talent of its own. This wait is not necessarily for a future that has been promised but has so often turned out to be false. Rather, she waits, and is vigilant, in the way that a person on a vigil watches over a dying person, not in the hopes of bringing that person back to life, but simply to accompany the body during its transformation. We might hope to do as much with the dead time in our lives.

SELF-PLAYING ART

Half of the fun of computers in the 1990s—for me, at least—was that they would break all the time. If you entered negative numbers into a program expecting only positive numbers, you might get it to regurgitate a mess of nonsensical symbols, crashing in spectacular ways. If you hit on the right parameters, you could get a program to spawn windows endlessly, until the computer ground to a halt, its beach ball spinning slower and slower until that too froze.[1] My delight in mishmash of glitchy windows and buggy code continued long after the industry got its act together and made things user-friendly. Indeed, in his history of "cool," Alan Liu argues that a feeling of subversiveness characterized computer culture in the 1990s and early 2000s, where the work itself was plain—in my case, I dipped in and out of university computer labs and anodyne corporate giants such as IBM—but the fun came out of finding ways of pulling apart those business systems.[2] In Liu's terms, such moments of "destructive creativity" allowed IT staff and other information workers like myself to experience a sense of individuality within the monotony of corporate culture.

I was only vaguely aware of the fact that other, more experienced pro-grammers and artists were taking these moments of "destructive cre-ativity" seriously, or as seriously as line art and blinking text could be. So-called net artists were finding ways to disrupt the web browsing expe-rience and corrupt data files, showing the web in all its glitchy glory. JODI (Joan Heemskerk and Dirk Paesmans), for example, modified (or, in the lingo, "modded") the first-person shooter game Quake to variously erase all characters from the game, alter character movement, and render a 3D scene as black-and-white op art. *Untitled Game* (1996–2001) led viewers to see and interact with a game with new eyes: it turned a game about shoot-ing people into psychedelia (figure 2.1). Like other affiliated "glitch artists" at the time, these artists of "destructive creativity" shared an affinity with tactical media and hacktivist groups, such as Critical Art Ensemble or Electronic Disturbance Theater, which used similar techniques—system errors and denial-of-service attacks, for example—to create electronic forms of political resistance.[3] They saw error, noise, or feedback as ways to create potentially liberatory channels of communication within an information system or to critique the seemingly smooth and frictionless functioning of those systems.[4]

But in what artists and critics have described as today's "post-Internet" era, a term that captures a sense of the Internet's newness receding into ordinariness, such work has lost its power.[5] Digital capitalism has made glitch art and hacktivism feel less subversive by co-opting their meth-ods.[6] Gaming a computer system might have once carved out a small space for autonomy within its rules, but gaming and hacking systems are now highly desirable qualities for potential employees.[7] And while the avant-garde media artist believed that introducing error or even viruses could feed back input from ordinary citizens into a centralized system,[8] today's digital platforms compel and even coerce feedback from even user, to the extent that critique itself has become ensnared in or flattened to something akin to product testing and quality assurance. Today's form of capitalism, political theorist Jodi Dean explains, "captures critique and resistance, formatting them as contributions to [its] circuits."[9]

Figure 2.1
JODI, *Untitled Game: Ctrl-9* (1996–2001), sound CD-ROM, http://untitled
-game.org.

A clear distinction between the user and the system of digital control
has all but eroded away. Net artist Guthrie Lonergan created a chart in
2007 to sum up this shift;[10] it describes the motto of Net Art 1.0, an aesthetic of "hacking" and "breaking," as "The Man is taking our privacy!,"
while the newer, more self-reflective motto, reflecting what Lonergan calls
an aesthetic of "banality," is "We willingly give up our privacy . . . why?"
Lonergan's description of a shift from outsider to insider rings particularly true for avant-garde media artists, who are positioned at digital
capitalism's bleeding edge. Their job closely resembles the flexible worker
who lives according to the "new spirit of capitalism."[11] Granting agencies
such as Creative Capital hold workshops on the topics of "key business
and management skills; strategic planning; goal setting; communications
and negotiations" to teach artists how to be entrepreneurial, to manage
subcontractors, to form alliances and connections with others in and
outside of the art world, to network, and to exist within an increasingly
networked economy.[12] And media artists' visual work overlaps uncomfort-
ably (though often deliberately) with billboards touting an iPhone user's
potential to create beautiful photographs, or with the user-"creators" of
Microsoft's operating system, "Windows Creators Update"—campaigns
that are less about visual aesthetics than signs of those companies' desires
to more closely integrate users into digital platforms that produce and
capture engagement, expression, and affective labor. When artists critique
the digital system, they simultaneously tend to use or circulate within
those same systems and platforms. The same difficulty applies to me: a
friend once pungently likened my work to "ambulance chasing" because
I rely on the subjects of my critique for my livelihood.

The entrenchment of these relationships means that new artworks are
far less celebratory. Critics have pointed to the way that contemporary
media art seems detached and marked with the "affectless *blague* of non-
criticality";[13] some are so reticent to speak out (or be critical) that they seem
to exemplify what art historian Hal Foster once described as a "'whatever'
artistic culture in keeping with a 'whatever' political culture."[14] In art critic
Joseph Henry's description, today's wave of "depressive Internet art"

consists of yawn-inducing, cynical artworks that offer "muted affect and shoulder-shrugging politics."[15] The "whatever" that they seem to produce may be too flat to register as an aesthetic experience (let alone a critical one) and too ordinary to appear as an affect. But many of these artworks go beyond pointlessness; rather than indifference, they exemplify a *lethargic* mode in art. Their reticence, ambivalence, and recessiveness of affect results from the position of being stuck inside a digital system that is all but impossible to escape.

For an example of this mode, take Cory Arcangel's *Self Playing Nintendo 64 NBA Courtside 2* (2011), which is a Nintendo 64 basketball game modded to make Shaquille O'Neal miss a free throw—over and over. From warming up to the missed shot to a reaction shot of Shaq's face, it is gameplay stuck on autopilot, endless and impassive (figure 2.2). As with another artwork he created the same year, *Various Self-Playing Bowling Games*, in which Arcangel modded bowling games to throw an endless series of gutter balls, this artwork embraces failure. In both, Arcangel signals his debt to the glitch art of the 1990s, such as JODI's *Untitled Game*. But there is something odd here about Arcangel's update, if it is an update. Rather than revealing anything new about a game's interface or logic to its player, *Self Playing Nintendo* seems intended to produce boredom or monotony. Nor does it reward a viewer for waiting and watching more of the loop; a viewer can simply walk away, with little consequence. Arcangel's non-interactive artwork is almost onanistic (or "self-playing") in its unproductivity, that is, in its failure to engage its viewer, or to offer a new interpretive context.

Whether a viewer is present or not, Arcangel's self-playing games play identically; an onlooker's participation changes nothing. The games work, in other words, without a subject. By making stagnant what would normally be a personal and interactive game experience, a viewer is briefly placed inside the first-person position of the game player, and as they continue to wait for "something" to happen, they are eventually left outside that experience of liveness. That link between viewer and player in the game slackens; the viewer walks away, presumably giving up on the

Figure 2.2
Cory Arcangel, *Self Playing Nintendo 64 NBA Courtside 2* (2011), hacked Nintendo 64 video game controller, Nintendo 64 game console, NBA Courtside 2 (featuring Kobe Bryant) game cartridge, single-channel video (installation view: *All the Small Things*, Herning Museum of Contemporary Art, Herning, Denmark, March 2014–September 2014). Photo: Sacha Maric. © Cory Arcangel. Image courtesy of the artist.

piece's potential for critical interest or pleasure. Though they can reenter the position of Shaquille O'Neal at any time, a viewer of *Self Playing* ends up witnessing not just the failure of their digital avatar to make a shot, but their own inability to align with the game player. While *Untitled Game* recontextualized a game for the purposes of critique, *Self-Playing Games* are instead cut-and-paste artworks that duplicate loops and procedures until the relation between viewer and subject becomes exhausted.

That failure is, however, the point. For digital capitalism functions by conflating those two terms that Arcangel cleaves apart, the "user" and the "self." It does so by appealing to our desires for expressive freedom, friendship, and sociality, and by obscuring the fact that a user account is simply, as it was at its origins, an accounting measure of computer time used.[16] This appeal also means that our relationship with our presence in data is fraught.[17] Even though digital platforms simply ask you to "be yourself," that datafied self that digital platforms ask us to identify with is not just an idealized, better, Instagrammable version of that self (though that may well be the case). On a deeper level, it is also a self that is never in perfect synchrony with the phenomenological present, but exists somewhere between past behavior and future projection, and somewhere between population and individual, as algorithms construct "your" preferences through the preferences of other people who supposedly think like you. Yet the bond between the subject and the user is tenuous at best; the user always threatens to regress into passivity (recall, for example, the drones of Apple's "1984" ad). For this reason, a variety of social and algorithmic incentives exist to nudge a nonuser back online. These inducements are mostly gentle, as when LinkedIn emails remind a disconnected user to log in by recounting a day or two of their connections' hyperkinetic activity, but user agreements also spell out penalties, such as account termination or inactivity fees, for not using enough.

Lethargy arises not just from exhaustion or being ill at ease, but also because digital platforms specify that only certain forms of interaction— most obviously, interactions that generate value—count as "being yourself." This is the case even when the value is not quantifiable as revenue, but simply what economists refer to as the network effect of having a critical mass of users interacting with each other (or simply watching or observing others), or in the data generated by observing consumers in the wild. But a vague feeling of disconnection, of being not quite "yourself," still lingers, even if only indirectly noticed as a recessive blockage in expression. This is paradoxically why art that fails to speak up—art that is reticent—is such a useful way to explore a history of the contemporary moment.

In these terms, Arcangel's *Self Playing Games* are transitional artworks that straddle two modes of feeling. They continue to have one foot in the subversive (and modernist) aesthetic of glitch and error, but they also begin to describe the lethargic mode that the artists I explore later in this chapter realize more fully. Whereas Arcangel's self-playing works are by design disembodied, even uncannily immortal in the eternal return of digital figures such as Shaquille O'Neal, those later artists make the body—and, by extension, its fatigue and its limitations—central to their practice. They reveal what art historian Christine Ross describes as a depressive "insufficiency of the self in relation to itself."[18] These insufficiencies produce an "aesthetics of disengagement," she writes, art that enacts the detachment of depression by producing the conditions of stunted or failed relationality. The implicit relation between viewer and object—traditionally understood through aesthetic judgment and discernment—is blunted by these artworks; they are designed to produce not critical engagement, but a state of disengagement.

Two pieces by the Dutch artist Liza May Post help Ross illustrate what she means. In *Bound* (1996), a figure is secreted within a mask that covers nearly all her face, and dressed entirely in a padded white uniform that, as Ross points out, makes her resemble the furniture around her. Turned away from the viewer and toward a mirror or console, the figure almost anticipates Arcangel's "self-playing" artwork that similarly preempts user interaction. In a video installation, Post's *While* (2001), confetti falls on immobile, seated figures; this action is broken only when a man briefly stands up to brush confetti off a neighbor, only to have this gesture of care ignored. Giving up, he returns to his seated position without establishing mutual contact. By placing the viewer inside the durational structure of a video loop that is never resolved or completed, *While* works differently from artworks that function using spectacular action; instead, it makes manifest a sense of unfulfillment, of time being passed and whiled away.

Such works of disengagement contrast with the socially reparative tendencies of the recent past, such as the "relational aesthetics" that aim to restore a sense of community to the public;[19] think of Rirkrit Tiravanija cooking Thai food for gallery viewers, for example, so they can socialize with their fellow eaters, or Tino Sehgal training docents to converse with viewers as they walk through the museum. These tendencies rely on intersubjectivity and communicative action, but, as Ross points out, they almost always rely on a certain "communicative competence," a consensus about what constitutes valid communication.[20] On these terms, a subject who doesn't subscribe to or fit this consensus—for example, a neurodiverse subject or, in Ross's book, a person with depression—is set up to fail. A subject disinclined to engage with others, for example, would be poorly suited for a relational artwork that implicitly requires certain types of social interaction, such as leaving the house to make small talk with wealthy strangers in an art gallery.

Rather than attempt to fix the problem of a broken social fabric, it's worth first understanding what we take for granted in posing the problem. Though Post's artworks are composed of non-events and refused events, they are not purposeless; they function, instead, as Ross writes, by "inserting the viewer into the devaluation of connectedness."[21] If connectedness has become the "infrastructural basis of everyday life" (Patrick Jagoda),[22] devaluing it explores a life where this infrastructure is missing, not taken for granted, or, even more radically, not wanted. Such an "aesthetics of disengagement" reveals the norms that undergird today's subject, such as communicative competence. These artworks show how the supposedly liberatory ideologies behind reparative artwork—intersubjectivity, relationality, conversation—assume and even reproduce normative ways of being. In short, they offer a scenario where we do not automatically aim to connect.

While a lethargic mode in art follows behind an aesthetics of disengagement, there is one important difference: the forms of lethargy may resemble the slow-moving and all-but-silent examples Ross draws from, but they may also look like the "busy idleness" that characterizes life on the Internet, such doomscrolling or compulsive refreshing.[23] Both depressive underperformance—that is, not communicating enough—and an anxious excess of action, even if ultimately self-defeating or mindless, can result from a user's sense of insufficiency. The oscillation between these two poles, and the variable field of sensations that result, is why lethargy is not simply a result of bodily tiredness or depressive withdrawal. Instead, lethargy is a response to a disconnect: the user lags or races ahead of the subject.

This disconnect affects more than just the individual; it also affects social life. The same conditions that create what Ross calls the "insufficiency of the self in relation to itself" also create, she argues, a "withering of the relational" in general.[24] The problem of how we relate to others that Ross highlights is particularly ripe for revisiting in a moment when we develop our sense of selfhood not simply in relation to other objects but also in relation to algorithms that generate value for shareholders or enforce state dictates. To be together with another person is also potentially to connect with their advertising sponsors or to misconnect with oneself in a confusing and messy space. Even as lethargy appears outwardly as the weakening of social bonds, lethargy is also an indicator that social life has also begun to change.

A BRIEF HISTORY OF DISENGAGEMENT, FROM COUCH POTATOES TO USERS

In 1976, as the story goes, Tom Iacino was on a phone call with another member of a Southern California humor group that favored watching television instead of pursuing a healthy lifestyle. He condensed the phrase "boob tube" and "tuber" into a tidy pun, giving birth to the phrase "couch potato." Another member of the group, Robert Armstrong, teamed up with a writer to pen the slim, pocketable *Official Couch Potato Handbook* (1983), which unreservedly embraces the stereotype of excessive television watching. It offers tongue-in-cheek anthems to television, television cuisine (toaster ovens figure prominently), advanced lessons on "simul-viewing" multiple TV channels at the same time, and a prehistory that connects television with Moses's fire, Plato's cave, and William Blake. What sets a couch potato apart from a casual viewer, the book claims, is his prolonged viewing sessions, which can make viewers "vegetal." To paint television as a device for turning viewers into vegetable-like objects is a running theme in *The Official Couch Potato Handbook*: at one point, the authors discuss watching sports until the viewer passes out; at another, they joke that it's hard to tell if a couch potato is dead or just glued to the tube. The handbook portrays its ideal viewer as devoid of agency—they have been "programmed," in a good way, by TV—and more object than subject.

These jokes about couch potatoes capture the zeitgeist of a country in the throes of economic stagnation; comics of Garfield, the lazy and overweight cat that enjoys eating pizza and watching TV, also appeared in newspapers starting in 1978. (By the mid-1980s, "slacker" had become a humorous insult.) As sociologist Alain Ehrenberg has argued, Western economies began to be based on an entrepreneurial model of personhood in that period: roughly put, the individual has "the duty . . . to *choose* one's own life."[25] Yet this entrepreneurial self, the model that underwrites today's user and its "duty to choose," came with a shadow, the couch potato. Couch potatoes get laughed at, and are supposed to get off the couch, but their laziness is also something to identify with or even—in contrast to a viewer's own sense of themselves as productive—to envy.

These jokes about couch potatoes are, however, ultimately about the laziness of white people: a reader of *The Official Couch Potato Handbook* quickly notices that there's only a single fleeting image of a person of color in the whole book, from a cartoon about unemployed couch potatoes in Detroit. In contrast, in the public discourse of the time, fear-mongering politicians invoked the supposed laziness of Black subjects—a long-standing trope of racism in the American context—as they railed against people taking advantage of the welfare system. So-called welfare "reform" efforts used this stereotype to argue that individuals who had otherwise been addicted to government welfare needed to be nudged into the workplace. Invoking the rhetoric of restoring agency and "personal responsibility," these efforts ultimately dismantled the welfare system through neoliberal economic reforms. This figure of sloth—typically an overweight, single mother glued to the couch instead of working—also extended an ugly American cultural trope of unruly and specifically Black appetites for food and sex. While this stereotype dates to nineteenth-century minstrel shows, similar performances after the Civil War used it to suggest that Black people, while officially free, could not be trusted with the responsibilities of self-governance.[26] Now the trope expanded to include the overconsumption of television. While the handbook makes the picture of a white population becoming fat and lazy humorous, the laughter has very little edge: during the same years, the specter of people taking advantage of the welfare system was roundly condemned. Whom we call a couch potato and whom we deride for taking advantage of the welfare system has less to do with actual consumption practices than with who has the privilege to be lazy.

While the couch potato was comically vegetal, the figure of a welfare recipient whiling away the hours watching television instead of working took on grimmer, almost-dead qualities: she seemed devitalized as a subject, only partially alive. It's an impression that reflects a widespread understanding of television as breeding an extreme degree of passivity. This belief, media scholar Michael Litwack explains, holds that "time spent watching TV is . . . time that is more dead than alive. Here, TV kills you, wastes time, wastes you."[27] Most immediately, Litwack refers to the link between the television viewer's sedentariness and obesity: the entwined histories of food and TV programming can be traced back at least to the 1950s, as in the frozen TV dinners invented in 1953. But he is also referring to the exemplary liveness that television seems to exhibit as a medium: it feels "live" because it focuses repeatedly on action over inaction, on a continuous flow of events (however contrived) rather than dead air. And that liveness, he argues, only serves to paper over the abandonment of populations that neoliberalism has disinvested in and ultimately left to die. While obesity is a feature of populations that have been marked for unhealth as a result of these structural conditions, obesity is instead commonly represented "as a crisis of agency and of the will"—of a lack of self-control in eating, of an inability to take action or to get a job—even of a passive viewer of TV. In this way, the lethargic subject becomes scapegoated as someone who refuses to engage themself.[28] Yet television's passivity has always been overrated; plenty of other mediums involve sitting on the couch and being sedentary, such as reading a book or being an audiophile.

Both the "couch potato" and the image of someone lazily enjoying their welfare benefits seem somewhat quaint today. We appear to have made a break from the medium of television itself and the viewership it engenders. In cultural critic Laurence Scott's words, we are in a "post-potato age"; even a user binging on a TV series online is likely to simultaneously check notifications on a phone or search the actor's bio in another window.[29] A user has become defined as a multitasker. (The 1996 epithet "mouse potato," describing a person hunched over at their computer mouse, never quite took off.) Even the political question around television-watching welfare recipients seems to have retreated into the background, having served its purpose now that the US welfare system has been dismantled.[30] Contemporary life is typically portrayed as an age of perpetual anxiety and frantic attention, of multitasking and nonstop notifications.[31] As media scholar Lisa Parks puts it, "whereas television viewers are perceived as passive couch potatoes, computer users are by comparison seen as active and overworked."[32] Algorithms prod us to engage and to be active in order to create ourselves as users.

Yet experts continue to warn us that interactivity can quickly become just another form of screen addiction. Apps for killing time are, in one critic's phrase, the digital equivalents of high-fructose corn syrup; they are giving us "cultural diabetes."[33] Indeed, many of the same metaphors of overconsumption remain;[34] the link between consuming food and consuming media reemerges in games such as Candy Crush and other bright, primary-colored, bite-sized games that game scholar Christopher Goetz describes as having a "fruit-snack aesthetic": they resemble "sugar cereal."[35] Similarly, technology magazines describe today's look of over-saturated colors that pop on a smartphone display—aesthetic decisions encoded into filters and computational photography, and then normal-ized as simply how photographs now look—as "eye candy." Critics even compare screen addictions to binge eating: "In today's world, technol-ogy may be more like food than it is like alcohol. . . . [Consumers] must rebuild their relationship with food while continuing to eat every day."[36] Public health scholars and journalists seek out guidance from the US Food and Drug Administration and the World Health Organization for "screen addiction," reasoning that the screen is like sugar and that sites like TikTok and Instagram prey on our cravings for it.[37]

Perhaps the main difference between the couch potato and the user is simply that it is almost impossible to opt out of the work of being a user. Users now interact even when they want to space out—that is, they now *actively* space out—meaning that the division between activity and passivity has become even more tenuous than before.[38] The algo-rithms that track user activity turn doing nothing into a sort of interaction that is captured as data: Netflix, for example, reads a failure to click on a suggestion as a lack of interest in it—a signal that informs their algo-rithms for future suggestions.[39] Algorithms can now infer users even in the absence of tracking cookies or explicit user accounts, for example, by using probabilistic measures to calculate the uniqueness of a user's browser identifier and IP address across websites. And the recent surge in platforms designed to monetize lurking, idleness, and casual viewership, such as Twitch, the video-game streaming site that Amazon bought for $1 billion in 2014, are a sign that so-called passive users are simply the next marketplace for digital capitalism's expansion; a user's interests can be easily inferred by examining whom they watch or follow.[40] Passivity is just another way to interact with the systems of digital capitalism.

The only people who seem to be less affected by this new reality are the well-off. Researchers found that children in more affluent families log on an average of ninety minutes a day less than low-income families; the phenomenon of Silicon Valley venture capitalists and executives who have banned or severely limited screen time for their children is so widespread that the *New York Times* has called it a "consensus."[41] Yet their employers and ventures promote screen time as a cheap substitute for public investment. In an underfunded school district near Wichita, Kansas, the Facebook-funded nonprofit Summit Learning attempted to "disrupt" education by distributing Chromebooks and implementing a web-based learning program that largely did away with interpersonal interactions.[42] The result of this experiment, as remote learners and workers quickly discovered during the COVID-19 pandemic, has been headaches, eye strain, stress, anxiety, "kids that all look like zombies," and even multiple seizures a day from one student suffering from epilepsy. To be off-screen is, in contrast, a relative privilege of those who can afford personal tutors, time off, or anonymity.

One group that embodies the seeming excesses of screen addiction is the Internet addict who doesn't want to work; today's equivalent of the couch potato, they are in government parlance referred to as NEETs (neither employed nor in education or training) or the "disconnected" youth. Political economist Nicholas Eberstadt estimates that there are roughly 8 million male members of the American labor force who stubbornly refuse to enter the labor force. In contrast to unemployed men who still hope for a job, he writes, "'socializing, relaxing and leisure' is a full-time occupation, accounting for 3,000 hours a year, much of this time in front of television or computer screens . . . Time-use surveys suggest that [NEETs] are almost entirely idle—helping out around the house less than unemployed men; caring for others less than employed women; volunteering and engaging in religious activities less than working men and women or unemployed men."[43] For Eberstadt, these screen-based activities seem to evince a bad domesticity, in which NEETs are too lazy to even "care for others," let alone enter the public sphere: "the death of work seems to mean also the death of civic engagement, community participation and voluntary association."[44]

Disconnected from the economy, the NEETs serve as a particularly galling irritant to political commentators on both the right and the left. To neoliberals on the right, represented by Eberstadt, this is because their on-screen activities seem to be neither work nor free time.[45] Free time is traditionally defined by the workday or work week, that is, as a way to unwind before the next day at the office or factory, or, alternately, as time off from the routines of housework. Consequently, watching TV, as those imagined to be exploiting welfare benefits supposedly did, or playing video games simply to pass the time, unmoored from any connection to work (waged or domestic), devalues the system of productivity on which both work and free time are predicated. This is why Eberstadt invokes the moral gravity of vagrancy: "Among those [men] who should be most capable of shouldering the burdens of civic responsibilities it instead encourages sloth, idleness, and vices perhaps more insidious."[46] To the left, NEETs seem antisocial or anti-communitarian, rather than prosocial, giving even left-leaning thinkers pause before championing their forms of idleness.[47] Yet this is because words like "work" or "community"— words that have come to seem intrinsically good for most of the political spectrum—have become, as scholars such as Kathi Weeks and Greg Goldberg argue, shorthand for institutions through which certain forms of responsible relationality and sociality are maintained.[48]

That the debate about screens is really about a debate over "responsible" forms of sociality is already anticipated by *The Official Couch Potato Handbook*. In one telling moment, the book briefly notes the existence of video games and newer, two-way television systems, but argues against interactive technologies: "Why watch television if you have to think and respond? If you are going to have to respond you might as well . . . cultivate friendships or read a book or something."[49] Its advice on intimacy is similarly asocial or lethargic: sex is "not recommended" as it would interfere with television viewing; in any case, "sex is overrated anyway."[50] And it describes television as a way of reducing crime: "When people are watching TV, they don't have time for antisocial activity," presumably because television makes them disengaged with society at large.[51] In these terms, the handbook offers advice that is unusually appropriate for today's networked age, where the new form of work involves "cultivating friendships" and "responding." Now, the new figure of unfitness, as it were, is not so much the couch potato as the user who is reticent, disconnected, or disinterested.

Accordingly, the forms of asociality represented by watching TV or wasting time on screens may be ways of coping with these pressures of connecting and being social. Rather than continuing in the moralizing vein of "sloth," let us instead turn to the South Asian word "timepass"—a word that, simply put, means doing nothing or wasting time. Sometimes used dismissively, to connote loitering or boredom, it is a refreshingly literal word that is distinct from the redemptive qualities of leisure. Television and media scholar Arvind Rajagopal tells a story of peanut-hawkers on a train who yell "Time-pass! Time-pass!" instead of "Peanuts! Peanuts!," because to think of a snack as timepass is to evacuate it as content, or even to evacuate it of the idea of eating, "referring instead to the duration of its value."[52] As with the meals sometimes served on transcontinental flights, the idea of timepass strips the gloss off travel, reminding us that such food is only incidentally for nutrition and largely to give passengers something to do with their time. The other preeminent vehicle for waiting today is, of course, media; no wonder airlines are increasingly replacing meals with in-flight entertainment.

As Rajagopal observes, timepass in the form of scheduled TV programming is particularly interesting because it can serve as a "time-sink," temporarily absorbing or deadening the flare-up of social or domestic conflicts (over, say, incompatible schedules) and thus providing a brief moment of privacy within an otherwise crowded home environment.[53] It imagines the television as a time-sink that cools domestic tensions and alleviates an excess of social obligations. Television is literally lethargic, a drug for shedding obligations, as long as one doesn't watch with too much interest, that is. It is even, as anthropologist Craig Jeffrey writes, a way of forgetting one's situation: "Avoiding negative introspection can become a job in itself."[54]

Timepass in a digital age offers a chance to temporarily forget the continual pressure for a user to connect and network with others. As Rajagopal's example of television as a time-sink suggests, timepass is a mode of waiting that is decoupled from proper sociality. Instead, it offers an attenuated, lower-commitment version of having to converse, connect, or interact with others, not to mention to continually account for oneself. In its digital context, timepass represents a set of forms, genres, and ways of coping that, in James Hodge's words, allow you to "be yourself without having to show up too much."[55] Some gestures, such as stroking your phone or compulsively tapping a screen, may seem mindless, but as Hodge, who studies new aesthetic forms of digital media, explains, they are key to "divest[ing] from the burden of subjecthood,"[56] because they enable us to temporarily retreat from specific moments of intersubjective communication to reinvest in the idea of relationality in general. Another of Hodge's examples is the way users capture a small dose of emotion in the reaction GIF, an animated GIF that contains a moment of joy, shock, sadness, or other emotion. The briefness of these forms has given rise to the phrase "the feels," which describes a small and often uncertain dose of feeling ("five feels equals one-third of a human emotion," calculated Katy Waldman in *Slate* in 2015),[57] a formulation that speaks to the ways of diminishing or containing affects from our own body without disconnecting completely.

Timepass is particularly needed for those for whom sociality can feel mandatory, rather than voluntary, such as the rideshare drivers who feel forced to be social with their passengers. Feminist performance studies scholar Summer Kim Lee has described this as a particular burden of minoritarian subjects, who are often asked to be "relatable" and accommodating to others. Lee contrasts such forms of enforced sociality with new forms of "staying in," such as an ode to masturbation by the poet

Ocean Vuong, or a music video that shows the singer Mitski alone and looking in on the outside of an All-American couple. Both examples might initially appear solipsistic, claustrophobic, or even narcissistic, but Lee argues that they are still part of a network of relations: "Their disclosures of the self occur through narratives of nonencounters with others. . . . They relate by the way of the asociality of the nonencounter as suspended, provisional moments of pause, hesitation, and rest."[58] Rather than walling oneself away from others, that is, the asocial subject continues to imagine relation with others, however deferred or postponed—a missed connection that is nevertheless felt as a connection.

Conditions of asociality, it turns out, are disproportionately, even overwhelmingly, felt not by the white male NEETs that Eberstadt describes but by people of color: in America, according to a 2020 report by the Social Science Research Council, a Native American person aged sixteen to twenty-four is 2.5 times as likely, a Black person 1.9 times as likely, and a Latinx person 40 percent more likely to be as "disconnected" from society as a white person of the same age group.[59] This suggests that the so-called crisis of asociality is an iteration of the same problem of gender and racial economic inequality that has been with us all along. For socialization is itself often a tool for excluding people. Lee's examples, drawn from Asian American culture, describe a racialized body that is always out of sync with the rhythms of sociality, who always arrives too late to the party (as in the Mitski video) or can never muster the reciprocity to be fully interactive. But Lee argues that this position of being "after" is also a position of unheralded potential. In her words, to be "'after you' is to create strange and estranging forms that shift how we understand and enact the critical interiority of Asian Americanness as standoffish, stalling, and a little if not wholly unrelatable, within what we deem to be meaningful forms of sociality and collectivity . . . [and] to insist on tardiness, slowness, lateness, and the problems for sociality and socializing."[60] That same phrase "after you" describes both the condition of a deferential server and also someone who always seems to ape, imitate, or take after the "authentic" identities claimed by whiteness, as in the Filipina clickworker I described in my introduction who models herself after a white American woman named Ashley Neves, or the stereotype of Chinese technologists as skilled at producing knock-offs of North American or European innovations.[61] But it is also a phrase that insists on talking about sociality through the lens of its temporality, and its resulting effects of deferral and lethargy.

Seeing lethargy as timepass allows us to give form to the otherwise null space in between the moments of interaction demanded of us, the moments when time is simply passed rather than being directed toward a purpose. If the typical impulse is to frame such time as a matter of moral concern, then we might reclaim timepass as a way of describing the intervals "in the meantime." Shilpa Phadke, a coauthor of a book on the right of women to loiter in Mumbai, has described her own online life in similar terms. Not utopian, she writes, but sometimes it's what's possible given the circumstances outside:

I wonder if I have the energy or enthusiasm to actually change my clothes and head out for that walk, safe or not. Often, I choose to loiter online instead. . . . Sometime I might have multiple chat windows open with friends loitering with me. I might also stop at a pornographic site or a sex chat site and wander in there, choosing to leave or stay at the click of my mouse. As I loiter, I could well be sipping a cup of *adrak* chai (ginger tea) or a glass of wine.[62]

That wandering is a combination of interest and disinterest, excitement and slack, leaving and staying: not the recipe for a revolution, but a way for us to while away time before we have to be ourselves again.

BUFFERING, CLICKS, AND BITS

To artist Katherine Behar, there is a silver lining to big data's seemingly inexhaustible appetite for amassing data about user behavior: "big data is . . . confusingly close to us and our bodies," she writes in her book *Bigger than You: Big Data and Obesity*.[63] Embracing this confusion—and also embracing the implicit metaphors of overconsumption inherent in the idea of "big data"—has the potential, she argues, to redraw and expand the borders between our supposedly sovereign individual selves and the users around us. One's data profile is often messy and indistinct, even inaccurate, because it always incorporates data about the crowd even as it simultaneously attempts to target an individual. By allowing the self to become more like an indistinct mass of data, and by accepting rather than resisting its objectification, Behar calls for a "more lethargic politics . . . not oppositional, not little, and not about action."[64]

What ideas go into a "more lethargic politics"? Behar draws on a tradition of feminist endurance or body artists to explore what happens to the body in an age of big data. She cites as influences artists such as Hannah Wilke and Orlan, performers who used their own bodies as raw material to be sculpted, cut, destroyed, or even disposed of. Orlan, for example, hired plastic surgeons to sculpt her face to resemble classical images of female beauty, modeling herself after Mona Lisa's forehead and the chin of Botticelli's Venus (among many others), while Wilke's self-objectification in photographs and video explored what it was like, in the words of one performance, "to be used, to be spread, to feel soft, to melt in your mouth."[65] While many of their contemporaries explored the contours of feminist subjectivity, these artists instead made their bodies into objects, even recipients of violence. That approach is key to Behar's own work, which accepts the body's objectification by big data and even makes that a starting point for exploration.

When we first spoke in 2016, Behar was in the middle of sourcing thousands of keys from a warehouse of surplus keyboards near Seattle, Washington; she was going to use them on the outside of a series of sculptures—one, a slumped data "cloud," referenced the cloud's etymological origins as *clod* or hill (this would become *Data Cloud (A Heap, A Mass, A Rock, A Hill)*, 2016), and another key-covered shape would become the basis of a performance, *Data's Entry* (2016), which choreographed a series of lethargic motions that would attempt to "enter" data. As we talked, a few questions kept coming up: How do our bodies register the presence of data? What kinds of new motions does big data produce? (We tossed around a few metaphors: perhaps big data was a type of clothing that feels too big, Behar suggested, in which one's body sloshes about.

Figure 2.3
Katherine Behar, *Data's Entry*, performance at the Pera Museum in Istanbul, September 2016. Courtesy of the Suna and Inan Kirac Foundation, Pera Museum.

Or perhaps the trail of data that one leaves behind is like the slime trail of a snail.) As opposed to the idealization of frictionless movement and flexibility that neoliberalism imagines—motions that are differentially available, as when populations are sorted into fast-track security lanes and holding camps near a border—Behar wanted to capture a sense of data as a material, everyday presence in the world "that we must contend with, as though it has a body of its own." We are weighed down by its accumulation, she suggests, and it is time we feel its corporeality.

Thus, in *Data's Entry*, performed at the Pera Museum in Istanbul, Behar had her dancers attempt, with vanishingly little success, to "enter" the world of data via a giant, elongated pill covered with keys, which she likened to an "unworkable interface"[66] (figure 2.3). Not only is the pill sealed and therefore un-enterable, it is also painted jet black, causing it to resemble the black boxes that cyberneticians theorize as the basic component of cybernetic logic. In this line of thought, computation necessarily occurs without users ever being able to access the inside the black box; for instance, software ideally should be able to store data without knowing what kind of physical storage it is (a hard drive, a USB stick, a cloud drive). The impossibility of accessing the interior of data forms the impossible task of the piece. But it is what data does to us—"data presses us, into grueling, repetitive tasks"—that forms the performance's conceit. As the title indicates, Behar is also thinking about the task of data entry itself—a mechanical task that is often outsourced to the Global South or to marginal populations, such as prisoners. An earlier video from Behar's *Disorientalism* (2005–present), an ongoing collaboration with Marianne Kim, suggests something of that dynamic by depicting the labor behind digital data. In it, Behar and Kim "appeared" inside a TV, frantically if repetitively tracing out the motions of Asian women assembling circuit boards with their "tiny little fingers" (as Behar put it to me, tongue firmly in cheek) to maintain and fix the microchips in the screen itself.[67]

Behar's choreography in *Data's Entry* captures a similar sense of laboriousness by requesting the maximum amount of effort and discomfort from her performers. In what we might call a reverse ergonomics, she is trying to produce fatigue, rather than eliminate it. "I give performers impossible tasks like 'Try to get your head and butt to touch,' and their struggles to operate within or around that directive are inventive in ways I can't predict. I always want to make effort visible," she told me. Behar also directed the dancers to move with an "excruciating" slowness, to further emphasize the theme of weight. The resulting attempts to connect with data are anything but flexible, smooth, or lively. A dancer dressed in white attempts to slowly contort her entire body into a pill shape to both resemble and surround the cylinder; her movements suggest data input that comes from not just the fingers but also the legs and arms. At times, the cylinder seems to bear down on the performer's back, making data a weighty burden. At one point, she uses her neck and chest to manipulate and hold it into place, attempting to let the pill "enter" her torso; elsewhere in the performance, she seems to embrace or even stroke the data interface, suggesting the strange erotics of the relationship.

A failed coup attempt scrambled my hopes of traveling to Istanbul to see the performance in person, so Behar thoughtfully sent me a video instead. As I continued to watch it in the following years, I realized that my early notes on the performance relied on a shaky assumption: that organic material is embodied and inorganic material is not. These assumptions began to bother me: Why do I see the loose clothes that the dancers wear as part of their bodies, and the plastic capsule as a separate entity? If that's the case, how do I explain the moments where the dancer and the capsule seem fused together into the same mass, as when they appear to be one person with two heads, or a person with arms and legs connected by a fifth limb? If I take Behar's claim that "big data is . . . confusingly close to us and our bodies" seriously, then I have to embrace the confusion, accepting that the "person" may be part object, too. So, yes, the capsule is a stone, a plinth, a weight, a burden—but also an appendage, an anchor, a device for respite, a companion, and a survival strategy. Still, though the capsule flirts with human form in the performance, there is no suggestion that it is coming to life as we conventionally think of it—fleshy, interior, active.[68] Instead, it suggests a devitalized form of life that is constituted out of data and the inorganic, and generated in the interplay between the data capsule and the performer. Like the couch potato and their television, personhood in *Data's Entry* is suspended between vitality and lifelessness. Personhood becomes lethargic.

After expending considerable effort to move slowly, the performer is exhausted, mimicking the gestures of resting even as they are directed to keep going. This is a stillness that is uneasy, tense, forceful.[69] This absence of action, however, is not meant to suggest a withdrawal from the data economy; it is not something like a strike or an act of refusal. As Behar told me, "The political act of doing nothing is about resisting . . . willful stasis, like Bartleby or [anarchist Nicola] Sacco." In contrast, she continued, "I think of doing nothing out of exhaustion as a mode of standing in place, reasserting oneself in place over time. It's persisting instead of resisting."[70]

Pointing to the fatigue of digital work, Behar moves away from the willful defiance of Herman Melville's story of Bartleby, a scrivener who "prefers not to" work or move despite the alternating demands and entreaties of his boss. Instead of laying claim to personal subjectivity, she emphasizes the value of persistence within a system.[71] We endure, and wait, and perhaps we are eventually unable to hold the pose her script calls for. In this moment, we become more like the plastic object that is material, exhausted, even disposable, returning a sense of limitation and even mortality to the deathless fantasies of the digital. In her words: "What happens when the impossibility of the interface is really tiring? In reality, you have carpal tunnel syndrome and sleep deprivation. But in fantasies of the digital, including digital work, no one ever gets tired."[72]

Digital work is not just manual labor, such as data entry; it also includes communicative labor. We post selfies of ourselves with our friends, or tag photos of our associates, but those @s and photo tags not only help facial recognition algorithms learn our identities and thus help governments to better surveil us, but also enable digital capitalism to profit from our expressivity. Undoubtedly thinking of this, Behar has developed a "Botox ethics," a reference to the cosmetic toxin that works by temporarily paralyzing the facial muscles that cause wrinkles to "create a 'dead zone' in the body."[73] While the face is typically idealized as the site of expressivity and subjectivity, Botox wipes out the face's ability to communicate, and thus its continual transmission of emotion to others—making it deadpan, as it were. By deactivating the face's communicative channel, a Botox ethics offers no way out of this regime of capture, but it might deprivilege the intersubjectivity and connectivity that for Behar are inexorably tied into neoliberalism's "compulsion to connect"—a compulsion that is then hijacked by surveillance algorithms.

Behar offers Botox ethics as an example of an ethical stance that she terms "necrophilia." Necrophilia, she argues, counters the "fetishization of liveliness" as well as the "imperative to network the self."[74] Here, she is simultaneously addressing neoliberalism's own lexicon of living as flexibility and networking,[75] and also a turn in philosophy and critical theory toward object-oriented ontology and new materialism, which, crudely put, attempts to explore the vitality, agency, and connectedness of non-human objects.[76] In contrast, an ethics of necrophilia, and its practices of deadpan and flattened voice ("deadvoice"),[77] would turn away from the constant need to express ourselves outwardly to others, often in order to "connect" with them and accrue value from that alliance. In an echo of the feminist art of endurance, Behar suggests that we inhabit the inward-looking, inexpressive status of the object, a practice of encountering others by "staying in."

We can see this realized in practice in *Modeling Big Data* (2014), when she positions a performer inside a large pink rubberized or plastic suit to model what she describes as an "obese data body, a body so stuck in the cycle of generating data that it is swollen and overcome by its own data glut."[78] The figure appears swollen to the point that a human figure inside is almost unrecognizable; the literal data-body appears so lethargic that it is barely able to move, producing a choreographed awkwardness, or a virtuous anti-virtuosity (figure 2.4). And its undulating motion resembles that of a natural organism, but the costume and mise-en-scene of the video again appear inorganic, plastic, recalling the well-known adage that "data is the plastic of [the] New Economy."[79] As with *Data's Entry*, this combination is disconcerting, because we expect body or data, and yet we see both at once. As scholar and critic Anna Watkins Fisher comments, "Behar turns our gaze to the liveness—and moreover, the creepy vitality—of big data."[80] Resembling a gigantic (if not fully erect) sex toy, it gives off a kind of tawdry corporeality that suggests a human part that has somehow become robotic in its function. Vitality here is shown not so much as a positive quality but as an unbearable extreme, as if it were impossible to escape an environment of liveliness.

In two of these performances, *Buffering* and *Clicks*, the struggle to move is particularly keen when one considers that both pieces resemble each other visually, despite their very different connotations—clicking

Figure 2.4
Katherine Behar, "Clicks" (frame from video) from *Modeling Big Data* (2014). Six-channel HD video installation, color, sound. Endless loops. Image courtesy of the artist.

normally signifying "activity," and buffering "waiting" or inactivity. The electronic soundtrack drives home this irony: while *Buffering* resounds with electronic drone and static, *Clicks* offers the more expected sound of a periodic pulse or ping, and yet the figure within it is all but unresponsive to the clicking sound. The result is to flatten activity and inactivity into a continuum: both are forms of idling or phatic activity,[81] the videos seem to say; we click around online as another form of vacating or idling our lives, as a form of timepass.

And though clicking ought to epitomize interactivity, from the computer's point of view (and perhaps the user's), clicking—indeed, interaction itself—is actually waiting for something to happen: we are always buffering, even when we are clicking. Because of the technology of time-sharing, a server's resources are shared with countless others and are subject to momentary (if usually infinitesimal) interruptions when another user is accessing the same resource. That is to say, most of the time the server or network is serving someone else—one of the thousand or million other people accessing the same resource at the same time, whether Google's search engine or one of Netflix's streaming videos. A movie-goer once spent half of the time in the theater looking at a dark screen, owing to the shutter that closes between analog film frames. If our time online were turned into a movie, we would spend upwards of 99 percent of that time in the dark—and yet perceive it as light.

It often feels that the world waits for us to interact with it—that the website waits for our order; that the smartphone waits for our touch. Yet it is this feeling of liveness that binds us to the rhythms of digital capitalism. Lethargy cleaves apart this sense of liveness, showing how the time of the user is full of barely concealed stagnation. And the couch in the background of Behar's videos becomes a waiting room, a symbol for how the time we call "live" is more dead than alive.

✻

In 2015, collaborators Tega Brain and Surya Mattu strapped a Fitbit fitness tracking device to a metronome, allowing a lazy consumer to fake a few thousand steps—in their data trail, at least—by merely flicking the pendulum of the metronome (figure 2.5). Brain and Mattu also produced satirical videos and pamphlets showing how Fitbits could be mounted on drills, bicycles, tires, and even pets, in a project they dubbed *Unfit Bits*. Why would one want to perform such an act of self-sabotage? For one, they write, health insurance companies have now begun to offer discounts for uploading Fitbit data to their servers, and a fitness spoofer could potentially save money.[82] And then there is the privacy argument: the sight of a private corporation tracking one's bodily motion and then converting that motion into dollars is potentially the start of a slippery slope; conceivably, a health maintenance organization could penalize someone for not moving enough.

Privacy is the most obvious frame for interpreting this work, but what should we make of the project's "unfitness"? The veritable embodiment of

Figure 2.5
Tega Brain and Surya Mattu, *Unfit Bits Metronome* (2015), metronome and
Fitbit. Courtesy of the artist.

the old saw that laziness "hurts nobody but yourself," such laziness offers
a witty response to the neoliberal injunction that one should "take care
of oneself"—that is, as an economic free agent, one is solely responsible
for one's own health. For, Brain and Mattu observe, the "healthy, active"
lifestyle is an economic privilege, one that is out of reach (and literally
unaffordable) to many working consumers, who may "lack sufficient time
for exercise or have limited access to sports facilities."[83] Or perhaps they
simply prefer an unhealthy lifestyle, in which case they may purchase a

"smoker's edition" of their movement-spoofing pendulum, which sus-pends a swinging Fitbit above a "classic glass ashtray."

By conjuring and updating the figure of the couch potato for the dig-ital age, *Unfit Bits* satirizes the tendency for critics to posit self-care as a solution to technological problems—particularly when self-care is simply defined as another technologized mode of surveillance. In contrast, the lethargic user here is self-sabotaging and therefore not "fighting back" against any system; the hack, as it were, is directed toward oneself. (We can recall here Behar's tongue-in-cheek description of Botox ethics, in which one administers to oneself "a little shot of death."[84]) That one might want to give up on one's health, capacity, or potential creates an interest-ing stumbling block for a market-based system of biopower designed to optimize life, because it assumes each subject wants life and liveliness.

Instead, *Unfit Bits*, as with other lethargic artworks, calls the very idea of a subject-as-user into question. What these projects show us is that lethargy is a momentary forgetting or letting go of the self as an individ-ual, bounded, sovereign subject—or in the terms used by digital media, as the user or a program's "executor." Though similar in outward appear-ance to escapism, and often overlapping with it, escapism may paradox-ically be more active: it constructs a specific world or an alternate role that one actively engages in or identifies with (a superhero character, a fantasy hero, a videogame avatar). Candy Crush, that repetitive, chirpy game we encountered earlier, is less escapist because it can be played for a few minutes and then put aside when the player loses interest. As timepass, it is designed to allow for or even to produce disinterest; the game industry considers it a so-called casual game because it can be eas-ily set aside. In media theorist Scott Richmond's take on that game, its production of what he terms "vulgar boredom" results from the "failure of the object to engage the interpretive, depressive I."[85] Though it differs from the critically interesting boredom of high modernist art (think War-hol's films here, where nothing happens for hours), vulgar boredom is useful because it allows us to be present to ourselves in different ways—for instance, by letting the mind wander—than when we are expected to respond aesthetically. As a scaling-back of intensity, as a form of timepass, lethargy attenuates the otherwise tightly coupled connection between subject and user. In turns, it briefly attenuates the coupling between the subject and the work of communication expected from them, without severing that connection altogether.

This reticence is what gets lethargic artworks labeled as apathetic and even cynical, particularly if we are used to artworks that explicitly declare their political intentions. But lethargy is a zero degree of politics that ultimately becomes a politics by other means.[86] Literary theorist Stephen Best has provocatively interpreted the Black radical tradition as coming not from "a desire to bring about positive social change" but instead from "violence 'turned inward.'"[87] Analyzing Cedric Robinson's examples of self-negation that "continue . . . to evade Western comprehension,"[88] Best describes the *palmaristas* in Brazil, settlements established in the seventeenth century by escaped enslaved people and their fellow travelers that they would then abandon and burn whenever the Portuguese approached, as a "society leaving no perceptible trace of itself."[89] Incomprehensible, but that illegibility—that evasion of Western comprehension—is its own politics. Because lethargy is similarly "turned inward," it is not easy to assess on the usual terms (e.g., Did it expose something hidden? Did it lead to economic gains? Did it resist a dominant system of power? Did it build community?). But acts that stay in or even wear away at the self may simply be what's possible for the digital underclass. And that "self-divestiture," in Best's words, not only offers some "fleeting relief from the pressure to endorse what Kant calls the world 'as is'"; it clears space for other modes of thought and even pleasures to emerge.[90] The reticence of self-divestiture is, then, the point; it makes communication the problem, even as it delays but ultimately abides the communicative act.

If lethargic art resembles a "whatever" artistic culture, then the word "whatever" is not meant as a gesture of subversion. Though Dean argues that the act of saying "whatever" is a "glitch in orality" that defiantly flings back the act of communication to the sender,[91] the lethargic subject does not rise to the level of irony, refusal, or insolence. Instead, lethargic art, as we saw in *Data's Entry*, replaces the idea of "doing nothing" as resistance with the questions of embodiment raised by "doing nothing," whether out of exhaustion, fatigue, race, or gender. Thus, the "whatever" present within lethargic art is less a glitch than the shrug of the teenage girl (the speaker in Dean's example) who knows she must do the task eventually—going along with the system, while deferring action for now.

"Whatever" is a temporary gesture, but perhaps, like the phatic "um" or "uh-huh," it is ultimately a way of reconciling those two irreconcilable temporal scales—the hyperattentive command logic of cybernetic communication, which demands a response in real time, and the limited capacity of human response. To offer a crude example, under the guise of user empowerment, my phone frequently presents me with notifications and alerts that seem to demand my immediate attention or resolution, emails that require my urgent feedback. Unable to respond and feeling unfit for interaction, I end up hitting the "snooze" or "dismiss" button, and instead waste time on Candy Crush or the equivalent. The alerts from friends and coworkers are still there, but after enough of them accumulate, I begin to see them as timepass: not as opportunities to connect or socialize, but simply a catalog of words to flip through, like a stack of old magazines.

And what if my indecisiveness were itself a choice? Choosing indecisiveness might send up the false "decisions" that digital culture is built from; the indecisiveness I feel is the feeling of impasse, of bad choices. If the act of connecting with others is *the* new form of work in today's digital environment, lethargy is what happens when connecting is greeted with indifference. For deferring is, ultimately, a way of creating slack in a temporal regime. Like the temporal structure of lethargy itself—ongoing and also interruptive—it turns away from the contemporary moment's always-on temporality. If a network produces a continual if low-grade sense of crisis by continually demanding interaction or input from a user,[92] lethargy helps us see this structure not as a crisis that can be solved by spectacular action but as an endemic pressure.

After all, indecision is a way of stretching out or prolonging a decision until it is no longer an inflection point with an abrupt before and after. Rather than opting out of the network entirely, lethargic actions such as indecision offer ways of remaining or persisting inside it, albeit in a devitalized state. In doing so, they redirect our attention away from the time of liveness and toward the "dead time" that constitutes it. By reframing time as something to be passed or killed, they remind us of the world that is both alive and also devitalized—a world in which persons are turned into objects—and the biopolitical mechanisms that bestow the temporality of liveliness on some and abandonment, withdrawal, and deadness on others.

The way a couch potato listens, receives, and defers action is its own form of agency. Here it seems fitting to give the last word to literary scholar Anne-Lise François, whose work on a theory of "recessive action"—the gesture, in Romantic novels and poetry, of "making nothing happen"—poses a quiet counterpoint to the temporal structures of interactivity. François retells a parable from Roland Barthes's *A Lover's Discourse*:

A mandarin fell in love with a courtesan. "I shall be yours," she told him, "when you have spent a hundred nights waiting for me, sitting on a stool, in my garden, beneath my window." But on the ninety-ninth night, the mandarin stood up, put his stool under his arm, and went away.[93]

"One waits, and waits," François writes, "and then gives up—such a movement yields a temporal sequence set loose from the ordering energies of the quest for possession and freed from the pendulum of anticipation and (non)fulfillment."[94] One could say the same about lethargic users: they wait, not because it helps them get ahead, but because to endure something is to mark time differently.

(HUMAN) ROBOTS

From across the Internet, it is not always easy to tell a human from a robot. In order to detect bots that can be used to create fraudulent accounts, scrape data from websites, or send spam emails, programmers have created tests known as CAPTCHAs (Completely Automated Public Turing test to tell Computers and Humans Apart), but they can be defeated. Spammers passed an earlier CAPTCHA test, involving entering the numbers in an image, by rerouting these images to poorly paid workers in Bangladesh or India, who would—almost instantaneously—return with the result; in 2013, a security firm created an algorithm to outsmart the numbers test, once and for all. Now Google, owner of the most popular such test, reCAPTCHA, uses a secret risk-analysis algorithm that measures how "reputable" and therefore how human your behavior is: what and how old your browser history is, how quickly you react, the time of day you tend to browse the Internet, possibly even whether you take breaks from browsing.[1] (Not surprisingly, it benefits when you fail, because then it forces you to teach its computers how to recognize cars, stop signs, and storefronts.[2]) You invisibly cross the line distinguishing humans from robots many times a day, and usually don't know you are doing so—until you do something a little too mechanically, and a prompt pops up, asking you to prove you are human.

These tests are behavioral, but how does one define what "human" behavior is? Search the web too quickly, and Google will challenge you to prove you are not scraping its results. Interact too slowly with a video—for example, by letting it autoplay in the background—and YouTube will ask you to show that you're still there. What is the "right" amount of interaction for a human? Computer scientists' working definitions of "human" are often exclusionary: some algorithms look upon web browsers that don't load images with suspicion, reasoning that robots typically forgo images to load pages faster—but persons with visual disabilities may also forgo images.[3]

While that may seem to be an unanticipated glitch, digital culture has often relied on exclusionary definitions of the human. As media scholar Jennifer Rhee writes about Masahiro Mori's original theory of the "uncanny valley," which attempted to describe the feeling of unease that results when robots become a little too human seeming, the human is narrowly defined around lines of health and ability. In a graph of the relative affinity humans have for various things, Mori locates an ill person below a healthy person (but, thankfully, comfortably above a zombie); later, he sees a prosthetic hand, however lifelike, as uncanny, "limp . . . cold," suggesting that a person with a disability compares negatively with humanoid robots.[4] Thus the idea that someone who is blind might register as a robot is not a design flaw, but instead a logical consequence of a computational culture that is ableist. As developers work to further design "humanness" into AI systems, they continue to perpetuate the biases inherent within that definition.

That tests for humans can be fraught with inequality is not surprising if we take a wider historical view, which is replete with many examples of distinguishing between human and nonhuman or subhuman. For example, during the 1980s, in a previous era of automation, Detroit auto workers blamed their Japanese counterparts for layoffs and stereo-typed Asians and Asian Americans as embodying the "robotic" for their almost nonhuman efficiency. This process is not limited to automation, however; it is deeply embedded in Western culture—so much so that, as critical race scholars point out, race was invented as a technology "for differentiating subjects from objects."[5] The writer and cultural theorist Sylvia Wynter has shown that Eurocentric culture has repeatedly fash-ioned "humanity" in its image by distancing itself from what it imag-ined to be nonhumans or subhumans.[6] Imperial cultures have crafted ever-evolving sets of legal, theological, and political distinctions over the last half-millennium to define colonial subjects as nonhumans and dis-possess them of their land and bodies, whether indigenous populations who had rejected God or Africans kidnapped into slavery or thought of as a "missing link" between animals and men. The modern if still pseu-doscientific equivalent of this is the theory of biological deficiency, made infamous by the 1994 book *The Bell Curve*, which suggested that certain races had lower intelligence quotients than others in part because their inferior genes propagate and accumulate over time.[7]

While this distinction has historically been policed by a central sover-eign, such as Wynter's theologians and humanists advising the Spanish crown, I have argued elsewhere that this power has diffused into a hybrid system of algorithms and users, which I term the *sovereignty of data*.[8] The colonial and color lines that exist on the Internet today for distin-guishing between human and subhuman are a product not just of cen-tralized ideologies but also of data-centric technologies. Thus, while the CAPTCHA test is nominally about defeating spammers and clickfarm-ing operations and media pirates, it is also a way of sorting people into categories because each category generates value differently. Even in the most literal of senses, a person who automatically passes Google's CAPT-CHA test without further requirements is a consumer, from whom data is extracted in the pursuit of marketing; a person who is marginal (e.g., a foreign visitor without a consumer profile) gets to recognize objects within images to train Google's AI engines; a person who fails (a "robot") is considered disposable, and is barred: other platforms will harvest their semi-mechanized labor.

And who are the robots? By using that word, I hope it's clear that I mean less the shiny metal robots of science fiction than the idea of a robot as a laborer. Its etymological root is *rabota* (Old Church Slavonic), which referred the number of days of compulsory labor required of serfs in the Austro-Hungarian empire, and now designates inhuman forms of work—regardless of whether the robot is made of silicon chips or cells. As servers that dwell just outside of or within imitations of human life, they also come from specific geographical locations. On the English-speaking Internet, they typically hail from the corners of the former American and British colonial territories owing to their English-language ability, such as the Philippines, Kenya, and Bangladesh:[9] though there is not a direct, one-to-one mapping between colonial powers and their former colonies in the pattern of labor markets today, we nevertheless see the continuity of coloniality at work. Scholar and activist Jodi Melamed writes that capital "can only accumulate by producing and moving through relations of severe inequality among human groups."[10] To take a closer look at the line between human and robot online, then, is about more than detecting fraud; it is to examine the uneasy junction of race, work, and geography in our digital environment.

The CAPTCHA test is perhaps the most explicit adjudicator of this line, but the idea of distinguishing between types of humans exists in many other places. Indeed, this type of discernment is what supports a racialized system of labor within digital culture, even when human bodies of color are visually or even physically absent. As we will see, distinguishing between humans and robots relies on what cultural theorist Sianne Ngai terms "animatedness," the old stereotype of racialized subjects as excessively or minimally emotional and expressive, and, simultaneously, puppet-like and lacking control over their own bodies, as if they had been animated by another.[11] The same logic drives racialized subjects to police their affect in public to avoid causing offense: playing up their "likability" or avoiding too much noise, or too much joy, as in Ralph Ellison's telling of the laughing barrels positioned in public places in the South, through which Black persons were obligated to shunt the sound of their laughter when they felt an urge to laugh.[12] Whether it involves seeing Asian Americans as repressed and unfeeling robots that are emotionally inscrutable, Mexican migrants as automatons for farm work, manufacturing, and construction (often literally—as in a sombrero-wearing automaton displayed at the 1938 Iowa state fair[13]), or Black people as hypersexualized and overly emotional bodies that are always out of control, animatedness becomes a technology through which subjects are racialized.

Ngai draws an especially interesting link between animatedness and media. Writing in 2005, she observes that the largest "live" events on television (O. J. Simpson, Rodney King, 9/11) all transmit the spectacle of racialized bodies. Liveliness, spontaneity, and zeal, Ngai speculates, draw their force from the liveness of the medium, allowing television to become a device to display and train us in racial difference. Just as television teaches its viewers to appreciate liveness (and thus liveliness) as immediacy and spontaneity, digital platforms teach us about liveness through varying forms of interactivity. Lethargy flattens this model. Built out of the "dead time" within daily existence, lethargy allows us to imagine life decoupled from liveliness and interaction. In this way, it raises the question of what it would mean to create moments of emotional recessiveness within a frenetic digital environment that continuously demands emotional feedback from its users. While digital capitalism classifies lethargic populations as something closer to robots than humans because of their supposedly flattened affect and their "canned" personality, this chapter explores what happens when lethargic subjects lean into their roboticness.

Consider one example of animatedness in meme culture. In 2018, thousands of fake accounts bearing gray, crudely drawn avatars suddenly appeared on Twitter. Though each "person" bore a biographical description, such as their aspirations for social justice, their creators—a group of alt-right trolls active on boards such as Reddit—also gave them expressionless faces and account names such as NPC4921337 (figure 3.1). NPC

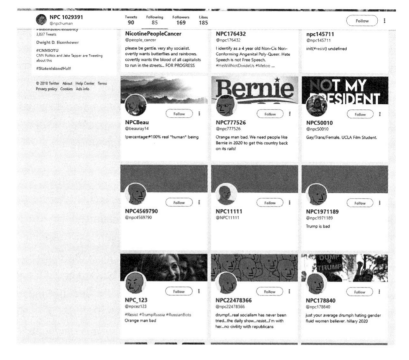

Figure 3.1

Screenshot of @npchuman's followers from Twitter, October 14, 2018. (@npchuman's account has since been suspended.)

is an abbreviation for "non-player character" and it refers to computer-generated characters in video games that are pre-programmed to interact with human players in certain ways, such as by saying canned lines about the weather. The trolls invoked the NPC's mechanical bearing and scripted responses to mock the supposed inability of liberals to think for themselves. In this rhetoric, the NPC-like liberals are "completely dependent on their programming," spouting the same one-liners about deposing Trump or killing capitalists as a result.[14] Indeed, the profiles' "About Me" section often contains pseudocode depicting their internal "programming": referring to Trump, one typical profile reads if(man. Color == Color.ORANGE) man.Bad = true;.[15]

However gray and colorless, these avatars were typically intended to represent (and belittle) persons of color: one avatar evoked the football player and activist Colin Kaepernick, while other avatars tweeted about "graycism" and used the hashtag #GrayLivesMatter.[16] Several meme creators also used NPC avatars to illustrate and represent the conspiracy theory, promoted by people like Kanye West, that Black people have been "brainwashed" by the media to vote Democratic.[17] Unable to think for themselves, not to mention emote properly, persons of color seem, in this formulation, to be blank robots that lack the supposed liveliness of the self-described free-thinkers. In this curious line of thought, the litmus test of one's humanity is a person's ability to "take a joke"—the joke being a racist or sexist one.

While this may be an extreme example, the Internet offers a myriad of other ways users test each other for animatedness. Consider the world of gaming, where the NPC meme originated. Examining the multiplayer game World of Warcraft, Lisa Nakamura, a scholar of race, gender, and digital studies, has demonstrated that many players profile certain styles of repetitive motion and stilted speech in gameplay as "Asian," due to the historical presence of other players based in China engaged in a practice of mining virtual gold for sale offline—so-called gold farmers.[18] Likening the gameplay of asocial or noninteracting characters to NPCs or nonhumans opens the door for everyone else to target them with vitriol, violence, or in-game death. Nakamura's example shows how a crude racial logic is reinscribed onto data bodies, despite the absence of visible signs of race; after all, every actor is an arbitrary animation on-screen. To make this distinction between human and nonhuman therefore requires some fumbling about, since there is no "naturally" human state: as Wendy Chun writes, the idea of the human is instead "constantly created through the jettisoning of the Asian and Asian American other as robotic."[19] (This is sometimes literally true: in one perverse case spotted by Katherine

Behar, a CAPTCHA test for humanness asked the user to type in the box the phrase "not from Asia."[20])

There is some irony here. By focusing entirely on the work of making money, gold farmers anger other players who are attached to an idea about gaming and expressivity or "fun": gold farmers aren't playing by the rules, paradoxically, because they are playing by the rules too well. In contrast, customers of gold farmers are wealthier players, often North American and European, who effectively outsource the repetitive work of mining gold to Asians so that they can focus on the more interesting or expressive parts of gameplay. This dynamic underscores a racialized system of labor that is at the heart of many digital platforms for "fun," sociality, or expressivity: liveliness is something that can be, and increasingly is, purchased from lower-wage microworkers elsewhere. Microworkers quench the seemingly insatiable demand by persons and corporations—as well as the social algorithms that evaluate them—for popularity, for having an audience, for being liked. They can encompass thousands of computers or phones and laborers, employed to stream a song, watch a video, or click "like" on a post in the thousands and millions; they amount to a vast transfer of "likability" from the Global South to the Global North.

These "like" factories are a rapidly growing part of the more general operations of outsourcing and reintegrating microwork. Computation itself has historically relied on outsourcing its labor: in the early part of the twentieth century, for example, the British Naval Office paid human computers—first clergymen, then young boys, then unmarried women or high-ranking women who couldn't publicly work—by the calculation at piecework rates (100 figures for 1 penny).[21] What's new about microwork is the ability for digital platforms to offer computational labor at piecework rates to anyone who wishes to hire them, and their increasing ability to make global labor interchangeable. These platforms, such as Samasource, Figure Eight/Crowdflower, and Microworkers, not to mention the better-known (and better-studied) Amazon Mechanical Turk, reroute computational labor to low-paid workers in the Global South (CAPTCHA-solving platforms are a kind of illicit parallel to these platforms). This effort is almost exclusively carried out for clients in the United States, Australia, Canada, and Western Europe, so they can do more enjoyable, creative, or higher-status work.[22] But their creativity or productivity comes about through the (computational) dehumanizing of microworkers: the same platforms typically conceal or "wrap" their microworkers inside what is known as an application programming interface (API). This means that microworkers become services behind the scenes, just like any other piece of code, so that a program can run a calculation or annotate an image without needing to know who or what—human or computer—is doing the work. To do so, the platforms must aggregate and anonymize the microworkers' output to make it as uniform as computer output, ensuring that these largely Asian workers will all "look the same." This anonymity has makes it harder for microworkers to take collective action, because a microworker typically has no ability to contact fellow microworkers (who may, in any case, be dispersed across several countries), let alone contest working conditions. And it also allows app developers, the digital labor researcher Lilly Irani tells us, to think of microworkers as mere servers, while thinking of themselves as creative users, "builders," and "architects."[23]

The digital platforms that position workers somewhere between animated and automated, the excessively human and the robotic, are largely invisible to end users. Wrapping humans in robotic form makes it hard to recognize their work at all, and it's rare for us to come across their traces: perhaps they live in the spam folder of our inbox, or in the image of a scanner's glove covering part of a digitized page on Google Books. The goal of that glove, as it is of data "cleaners" more generally, is to leave us with an image of data as born through a kind of immaculate conception. But if we look for racism on the Internet only in the visual depiction of bodies or results from algorithms, we miss one of the basic logics of how race on the Internet operates. Animatedness does not often show up in explicit ways (as in Ngai's example of television); it will likely not manifest as racist search results or on online forums. Instead, it is embedded within the infrastructure of labor that is deliberately erased from the picture, so much so that software developers themselves may not even notice they are relying on human labor.

Though racialized, the labor of human robots is imprecisely described by the metaphors we currently use to understand exploitation. Digital scholars often use the images of the "digital sweatshop" and the "maquiladora" to describe the labor system at the base of digital capitalism.[24] However sympathetic I am to those ideas, the lens of exploitation typically assigns most of the agency to malignant forces in the West (in Silicon Valley, in Washington), ironically (re-)erasing the agency of the populations that this critique purports to help. So it's an open question of how the microworkers themselves might view and work within this labor system, or of how well those metaphors apply to the ordinary experience of digital environments. Anthropologist Mary Gray's interviews with the on-demand laborers often pejoratively referred to as clickworkers showed that her subjects understand their job as one that "takes quite a bit of creativity and insight and judgment," rather than as a series of menial, sweatshop-like tasks.[25] And while scholars of digital labor tend to think only of microwork as clicking on images, moderating content, cleaning datasets, or converting images to text, these same platforms are increasingly used for the assembly and recirculation of emotional labor—even, as I detail shortly, as the labor of being animated isn't cleanly or endlessly extractable from the subject as computational piecework.

These platforms include the thriving field of "emotional AI," run by companies working to identify your mood—testing if, for example, you are really engaged with an advertisement or if you are simply feigning interest. These companies have collected vast databases of facial expressions and are now able to decide if your smile denotes a "puzzled smile" or a "crying laugh." These companies compile their databases from both staged photographs and facial images "in the wild," for instance, on YouTube. And human-cleaned datasets are invariably adjudged to be more valuable: existing algorithms for matching facial shapes to, say, smiles or frowns are not particularly precise and are often based on their North American and European programmers' idea of what emotion looks like.

Human "annotators," so named because they must annotate each image with its emotional state, must distinguish between closely related states, such as "pleased" and "glad," or "aroused" and "intense." To amass the large databases, which can span up to 450,000 to even millions of faces, requires an astonishing amount of labor, typically hired through crowdsourcing platforms to save money. The Microworkers platform even has a preexisting template for "emotional responses to image" and "image emotion evaluation." These microworkers are teaching and classifying emotions—or, in the terms that I have laid out so far, teaching animatedness—to tomorrow's robots. Even as new artificial-intelligence techniques attempt to produce robots under the theory that what defines the properly "human" subject is its emotional intelligence, they rely on what Atanasoski and Vora term "surrogate humans"—human workers that exist in a subaltern relationship to their masters.[26]

Samasource, one of the largest suppliers of human-cleaned data for artificial intelligence systems, describes what they do as bringing the humanness to AI—"Training data is the soul of your AI," as their business case reads—and simultaneously bringing a "better life" to East Africa, where most of their data cleaners work. (Its founder justifies its low salaries by citing the "humanitarian" reason of not distorting the local market economy.) But this emancipatory mission, Atanasoski and Vora insist, assumes a *racialized aspiration for proper humanity* in the post-Enlightenment era"—one where the microworkers serve as human surrogates for their white clients, appealing to a postracial model of liberal universalism but hiding the colonial lines that undergird it from sight.[27] The Baroque architectures for locating domestic work within the house of but apart from the master's quarters now take place digitally, where the "soul" of a home device that speaks from your bedside table has been animated by persons located in Kenya, persons who are marked as almost (but not quite) properly human.

Other applications of surrogate humanity include full-blown "technologies of care," as artist Elisa Giardina Papa puts it, such as a distributed team of writers providing a virtual boyfriend experience.[28] (These services, which create the fiction that one is in a relationship, are popular for a variety of reasons, including fitting better into one's social circle, not feeling left out, or simply for emotional companionship.[29]) On a digital platform, care need not be extensive, therapeutic, or even as personalized as customer service; as with other microwork tasks, it can also be made smaller and even banal, such as being employed to stand in a crowd before a politician's speech. One worker in Greece whom Giardina Papa interviews describes how her clients—presumably aspiring singers or stars—often hire her to post comments such as "U sing too good!," "U are so sexy," "nice pic," or simply to post emoticons on their Instagram feed. These posts may seem fake or false, and when we notice these "artificial" likes or artificial reviews, we typically react with disapproval or disgust.

In their seeming lack of authenticity, in their racialized otherness (like call center workers, microworkers are often required to pass as white Americans or Europeans), these workers seem to provoke more anger than most other workers we encounter daily; they are even exposed to the risk of violence by so-called digital vigilantes and "scam baiters," who aim to turn the table on fraudsters and subject them to mob justice.[30] Yet a telemarketer or microworker's lack of agency and authenticity—necessitated, naturally, by their job—may be the most interesting territory to explore, as it helps point to a wider and shared condition of lethargy. The type of digital lethargy this chapter explores is the affective state of being enframed by this supply chain, and consequently of being unable to do nothing. From the outside, lethargy looks like being more robotic (or, conversely, seeming less human). It can be either being less animated than a proper human or, paradoxically, being overly animated, because in the latter case, the suggestion is that one is animated from without.

To examine the lethargy within microwork, I turn to an artwork, *Canned Laughter* (*Risas enlatadas*) (2009), by Mexico City–based artist Yoshua Okón, which depicts a fictitious maquiladora in Ciudad Juárez. Instead of processing textiles or electronics, however, it manufactures shiny red cans of laughter destined for US sitcoms—cans labeled evil laughter, manly laughter, and sexy laughter, among others. A dystopian world where low-wage workers across the border, or around the world, laugh, cry, or otherwise emote for white audiences is not as far away as we might think.

CANNED LAUGHTER

In *Canned Laughter*, you enter a concrete space that resembles a factory space—indeed, in its first iteration, Okón rented a former assembly plant in Juárez—and come across a long table styled as an assembly line. On it, there are the cans, which you can listen to, and there are 1990s-style televisions playing video loops of corporate propaganda from the fictitious factory owner (named "Bergson," after the French philosopher who wrote about, among other things, laughter) (figures 3.2a and 3.2b). Hung on racks are workers' uniforms, and on the wall, there is a video loop showing a German conductor coaxing a chorus of Mexican workers through various types of laughter. "This laugh, a witch's laugh, is from my hometown, the Black Forest," he says at one point. A few minutes later, the video cuts to show workers at the assembly line, fictitiously injecting sound in the cans by operating a machine that dips a rod into the metal, or fictitiously testing quality by seeming to inspect the sound. The people in the chorus and on the assembly line are themselves former maquiladora workers performing their old roles, but this time for art.

How are we to interpret this? We might begin with the artist himself, who writes that *Canned Laughter* shows the "impossibility [of] translat[ing] and reproduc[ing] true emotions through technological means."[31] While Okón is clearly taking a shot at the fakeness of canned laughter, this explanation feels inadequate; what, after all, is "true" emotion? To suggest that there is a "true emotion" in the body that technology then distorts is to ignore the ways that the body is itself physically and culturally technical: love letters are produced by hands grasping pens and using alphabets and scripts to praise the beloved. Technologies mask, amplify, and convey emotion to us. And crucially, technologies shape the terms by which the emotion of others is registered, turning, for example, a non-response

Figure 3.2a
Frame from Yoshua Okón, *Canned Laughter* (2009), projection video in installation, courtesy of the artist.

by a gold farmer into an example of a racialized threat to the rules of the game.

Rather than reading Okón's artwork as humanist satire, we might be better off taking the artwork literally, that is, as a documentary on the very *immanence* of "reproducing true emotions through technological means." If it seems counterintuitive to call this essentially arduous and manual work technological, the difference is that rather than work progressing from human to robot, as futurists might have expected, we have entered a phase where it's more cost-effective to hire human robots—what Amazon founder Jeff Bezos terms "artificial artificial intelligence."

The artist's own description for it is the *infomaquila*, which briefly flashes on screen during one of the inane corporate videos that loops

Figure 3.2b
Yoshua Okón, *Canned Laughter* (2009), installation view, *Yoshua Okón: 2007–2010*, Yerba Buena Center for the Arts, 2010. Courtesy Yerba Buena Center for the Arts. Photograph by J. W. White.

on the television sets (figure 3.3). This is a word that he likely borrowed from the filmmaker Alex Rivera, who in his contemporaneous film *Sleep Dealer* (2008) describes an info-maquiladora in Tijuana called Cybracero Systems, Inc. (figure 3.4). A cybracero, in Rivera's description, is a cybernetic update of the 1950s Mexican guest worker, or *bracero*, tasked to pick crops for the US agriculture industry and then return home at the end of the picking season. The cybracero is "safely" contained behind a US-Mexico border wall; physically implanted with telepresence technologies, cybraceros animate robots north of the border to water lawns or construct skyscrapers, or, in a different form of (re-)animation, provide

Figure 3.3
Frame from Yoshua Okón, *Canned Laughter* (2009), monitor video in installation, courtesy of the artist.

"memories"—narrated feelings and impressions—to willing buyers. The cybracero helps to whitewash this future by moving Brown bodies out of view, just as today's microwork platforms help to mask the identities of digital laborers from their employers.

Exclaims one of *Sleep Dealer*'s characters, "They want work without the worker." This future was already set in motion when outsourcing firms in Mexico discovered that deportees from the United States had excellent American accents gained after time spent across the border: deportation creates the perfect transborder workforce. The laughter of Okón's own outsourcing firm represents a further displacement of national origin and, in turn, the racial capitalism underpinning this strategy. Imagining the workings of the fictional world Okón creates, Anca Parvulescu,

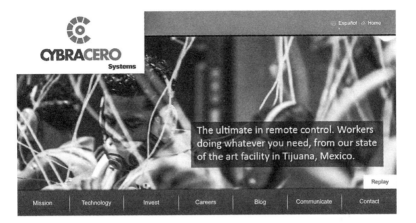

Figure 3.4
Screenshot from Alex Rivera, *Cybracero Systems* (www.cybracero.com) (2009), courtesy of the artist.

author of a cultural history of laughter, writes that the workers' laughter is "exported to the rest of the world, including presumably back to Mexico . . . [and] likely to be consumed by the same workers who participated in Okón's installation, after their working hours in the maquiladora."[32] Parvulescu's suggestion is that workers, if real, would be doubly exploited: they presumably do not recognize the sound of their own laughter on the sitcoms, even as they consume those shows to unwind after a day of work. They pay both times—with their body in the factory and with their eyes at night—and thus it is a form of "express[ing] their gratitude for their chance at being oppressed."

Again, however, I want to move beyond simply adopting and applying a model of oppression, which has been the lens through which each critic has viewed this artwork, and also the lens through which critics view microwork. The artwork's reflexiveness pushes us to talk about the *infomaquila* model differently, for it stages a problem about critique by making its ostensible message a little too obvious. Witness *San Francisco Chronicle* art critic Kenneth Baker, who writes that there is "no doubt as to Okón's intervention. Intended as critical satire of the global corporate order, it feels forced in every respect, at best pleasing viewers with the thought of the artist having employed some needy people for a while."[33]

To be sure, the literalism of seeing former maquiladora workers mechanically performing maquiladora work does lead to a sense that something, in the critic's words, "feels forced," as if someone telling a joke had insisted on asking, "Get it?" afterward. But what the critic misses is that the feeling of frustrated satire is part of the artwork's very design: the word *enlatada* means "canned," but as Amy Sara Carroll, a scholar of Mexican contemporary art, notes, it also means "canceled" or "put on ice," as in the English expression "the project got canned."[34] In other words, *Risas enlatadas* tells you, rather directly, that it won't succeed: laughter, quite literally, is canceled here. If one of the contracts of performance is that a performer performs, and the audience reacts, here, Okón severely limits the audience's ability to react: the workers seem to laugh on our behalf, as if they were a reaction GIF, *our* reaction GIF. This use of laughter short-circuits our response, making it all but impossible to resolve our experience through aesthetic judgment. Consequently, the artwork doesn't feel like critique or satire, nor does it allow much unguarded pleasure or humor. Instead, it moves us toward something more ambivalent: a feeling of expecting laughter to happen, and being disappointed, perhaps; a feeling of things being "forced," somehow, whether the satire itself or the laughter the workers themselves perform.

And the critic's dismissal of the artwork as a vehicle for "pleasing viewers" by employing "needy people for a while" inadvertently reveals one of the conceits of the project: if you are the pleased viewer, you are indirectly those workers' employer. As you look around, perhaps "sampling" a can of laughter or examining the rows of uniforms for their symmetry, are you not behaving like a factory supervisor? Perhaps you consider the assembly line, the ways workers are organized, and the logistics of the supply chain, as if you were an administrator of a process or institution.[35] Or perhaps you focus on the glistening, jewel-red cans (a clear reference to Warhol's soup cans), like a client coming to inspect their purchase.

None of these positions allows for easy opposition. This entanglement even suggests that American critics may be eager to read "exploitation" in *Canned Laughter* or in clickfarming platforms because it eases their own discomfort at being caught, a bit like middle management, in between the corporate structures of capitalism and the laborers in the Global South.

Exploitation may be an expedient way of describing the relationship between worker and work, but it doesn't account for other affective and psychic valences within and beside it. Exploitation tends to be deployed by Western scholars paternally, and presumes a passive and instrumental Global South. In reality, writes sociologist Antonio Casilli (quoting Payal Arora), those "'at the bottom of the data pyramid' are just about as involved in creativity, online recreation, and leisure—and just about as subject to mechanisms of data extraction" as users elsewhere.[36] There is, in other words, a more ambivalent relationship between the microworker and their work than outsourcer and exploited victim, and this relationship perhaps shows both the potential and the limits of the maquiladora metaphor for describing digital work. For example, even the most liberal-minded scholars tend to react poorly to much of the work of spamming or creating fake accounts or upvoting a post—there is invariably a moment of barely concealed distaste when, for example, they mention that clickfarming is not illegal in many countries. Yet in an analogous example from the better-studied industry of call centers, anthropologist Purnima Mankekar has shown that Indian employees often become call center agents out of a fascination with Western media, and call centers, often built like Hollywood sets, train their employees to act. The act of adopting another accent, or substituting one kind of laughter for another, need not simply be about impersonation, but must also consider desire, aspiration, and, to use Mankekar's words, personation.[37]

Similarly, microworkers often do not see their work as menial; it is instead a source of pride. The interviewees in Garrett Bradley's short documentary about microworkers in Dhaka, *Like* (2016), for example, describe their work as online marketing. They see clickfarming as a way of getting their foot in the door in the information technology sector, a creative and even cool way of finding work in a constrained job market.[38] In the words of one subject, clickfarming is a way "for maintaining, for surviving." Another subject points out that Facebook makes social connection seem like love; for this reason, he says, a clickfarmer "is treating it like prostitution: 'you'll pay me; I'll give you love or like.'" This is particularly trenchant, given how much of the exploitation in digital labor (indeed, as I have argued, the very subject position of the personal "user") has resulted from the mystification of economic framework as doing things "out of love." This is how online platforms convince us to give our time freely to write online reviews, and how Facebook convinces us to "like" the advertising posts inadvertently promoted by our friends. And it is also the line of thinking that justifies the low wages paid to Amazon Mechanical Turk (MTurk) workers, who are—in this line of thinking—housewives doing MTurk tasks as a fun hobby, in their free time, for a few extra dollars. (The platform positions itself as an enjoyable alternative to "just sitting there," but the recruitment of couch potatoes into the ranks of the gig economy is another instance of digital capitalism's clawback of idleness.[39])

This is not to dismiss the fact that microworkers and "info-maquiladora" workers may labor under poor working conditions—in extreme cases, rooms so crammed full of computers that the heat from so many processors can burn the skin. Because "sweatshops" are typically closed to academic research, we may hear more in-depth studies of benign workplaces than not. I simply mean to say that the very description of work as menial or robotic is troublesome. Recall that white World of Warcraft players often see gold farmers as inanimate non-player characters because even though the farmers are working all the time, they do not consider them active, agentive participants in the game. Indeed, ethnic studies scholar Iyko Day has shown that the supposed "excessive efficiency" of Asian workers, and their alignment with the abstraction of money (as opposed to the supposedly natural qualities of "concrete" work), is at the root of anti-Asian racism.[40] Analogously, we must be incredibly careful not to describe these workers as passive, robot-like laborers who could become *more human* by more dignified work—or by attention and description from Western saviors—since, of course, they are already human.

To supplement traditional models of digital labor, we might try to offer finer-grained descriptions of how one makes do within the compromised (digital) environment—including how one "maintains" and "survives," to quote the worker in the Bradley video, rather than (or in addition to) resisting. After all, the workers are caught between a rock and a hard place: of being forced to engage in repetitive, robot-like actions for work, for example, and yet needing to prove that they are agentive and human; of being required to express themselves while not having the ability, within the framework of animatedness, to do nothing. The ambivalent affective field that results is difficult to reconcile with theories that portray platforms as extracting affect from bodies. For one, affect is sticky: whether as a negative or positive, affect lingers after a transaction is over. And the intensities (or lack thereof) of feeling are hard to parse into real or manufactured, human or robotic emotion.

What *Canned Laughter* seems to show us is the laughter of the workers delinked from their affective states; they laugh to a rhythm demanded by a conductor, and it could be funny or not. "Funny" may coincide with the sound of laughter, or it may lag. In a model of exploitation, we might assume that all the laughter Okón's workers produce is "false," equating performance with inauthenticity. But a closer examination of the installation shows something else going on.

PERFORMANCE UNDER WORKING CONDITIONS

Consider a moment from the longest sequence of Okón's video, when the maquiladora workers are performing as a chorus; they stand shoulder to shoulder in their uniforms and are orchestrated by a conductor, while a sound engineer records on the side. As they laugh, they are also listening to themselves laugh, and this causes a slight out-of-sync quality to the sound, an effect underscored by the camera panning over individual faces that react differently and move differently to form the sounds. The sounds coming out of their mouths are coded as laughter, but they aren't necessarily moved by it: their faces articulate the canned phoniness of the performance. This leads, at times, to moments of meta-laughter: there is a woman on the assembly line who begins laughing spontaneously at the men in the chorus, presumably in an off-script moment when she is supposed to be silent. And a few workers occasionally break character, as it were, and seem to guffaw at the ridiculousness of the sounds they are producing—even as they are producing them.

One way of describing this moment might be to say that the woman and the man are laughing "authentically" in the middle of the task of faking laughter or, conversely, laughing "insincerely" as an attempt to resist the imposition of laughter. But rather than continually attempt to distinguish between authentic and fake, as the CAPTCHA test purports to do, consider that fake laughter can lead to real laughter and vice versa; indeed, laughter is inherently unstable. The formal composition of the video sequence records some of that complexity. Okón's camera cuts between two types of shots: first, a static wide shot of the conductor and his chorus, and second, a medium close-up of the men's faces. While the latter pans over a range of indeterminate expressions, the wide shot offers an overview of the overall system of organization, showing how the conductor produces a single product out of disparate voices. Even individual moments of foot-dragging or reticence that an individual worker may engage in are ultimately synthesized into a uniform product: this, after all, is the point of supply-chain capitalism, which factors in (and works around) temporary points of unreliability.

Nevertheless, the camera shows that these two layers of information form a circuit, and what this indicates to us is that canned laughter is sometimes produced by live, human workers, and, reciprocally, the feeling of liveness (the "performance," as it were) might be produced out of prefabricated feelings, such as canned laughter. As a result, the alternation between the two camera distances doesn't show a sharp cleavage between, say, the authentic individual and the fake product. Instead, the laughter indexes a state of emotional undecidability that occurs within this logistical system, where it is unclear what (or who) is feeling. Laughter, after all, is commonly produced in response to social and power dynamics. Particularly in the awkward encounters between two parties with asymmetrical amounts of power—say, art viewer and worker—one often laughs to cover over something else, over what we might call the friction in that difference: indeed, we might conclude that laughter is not just the cover for that ambivalence but the very embodiment of that friction.[41]

What I am claiming is that the laughter in the video operates at the unstable border between emotion and affect, a fuzzy hinge that is intimately understood by nearly every person who has written "lol" in a text message. "LOL," laugh out loud, rarely indexes actual full-on laughter; it is instead a marker that simply acknowledges the genre of humor. As the most popular definition of "lol" in Urban Dictionary puts it: "nobody laughs out loud when they say it. In fact, they probably don't even give a shit about what you just wrote. More accurately, the acronym 'lol' should be redefined as 'Lack of laughter.'"[42] The same thing goes for "haha": explains one commenter, haha "doesn't really reflect how much they are actually laughing/how much they actually thought something you said was funny." Both words are phatic gestures, like "uh-huh," meant simply to acknowledge that you are still communicating—indeed, even a way of acknowledging the need for you to respond, to be just animated enough.

"Lol" is typically not a masquerade, a way of cloaking one's identity from surveillance or data platforms, or pretending to be otherwise while maintaining a hidden inner life, all of which would require us to resurrect the distinction between fake (and therefore subversive and presumably agentive) and real (and therefore presumably not subversive and non-agentive). All laughter is at some level real. Instead, "lol" stands primarily for a moment of undecidability and *unresolved affect*: it asks for time to think or feel, akin to a "pause game" feature. "Lol" is not the full feeling of humor but rather the space or hesitation of it, the attempt to express something while not yet being ready to commit to a specific emotional end point; it is not what we could call "live" laughter but canned, underperformed laughter: laughter as "deadvoice."[43] In a moment where expression has increasingly become the new arena of work, such deferral is a form of lethargy, one of many "recessive actions" in which the subject aims to defer or set aside the burden of actualizing the online self. To recall the words of Bradley's interview subject, it is a way "for maintaining, for surviving." Writing "lol," not laughing but writing it, acknowledges the demand to reciprocate by acknowledging the genre of the demand; it maintains the connection with the requestor, while delaying the affect's stabilization for later.

I say "delaying" because outright refusal may not be possible in the always-on world foreshadowed by *Canned Laughter*. (Lethargy approaches but is ultimately not passivity, because passivity, like rest or downtime, is generally foreclosed to the lethargic subject.) Consider a final sequence from the video, where the workers are shown in the rather un-Fordist position of standing in a circle outside, holding hands and taking a company-mandated meditation break. As Okón puts it in an interview, they have been brought out so that they can be "thankful for being exploited."[44] The workers he interviewed said these breaks were common inside real maquiladoras, which have begun to operationalize silence and mindfulness as tools of productivity. This is arguably even more depressing than, say, simply overworking the workers: capital metaphorically captures all expression, even the absence of expression. And this operationalizing of silence is increasingly common in digital culture: a variety of techniques capture what a user fails to click on in recommendation algorithms, sell silence as a luxury that can be purchased, or use silence to profile a user's behavior.

Capturing the act of doing nothing points to a subtle but important shift. Traditionally, doing nothing is synonymous with idleness or vagrancy—that is, terms that are opposed to work—and they were codified in anti-vagrancy laws that deprived people of color, the poor, and sexual minorities of the ability to "do nothing." The United States imported these legal strictures from England, where they date back to at least the sixteenth century, and innovated upon it: Southern and Western states developed novel anti-vagrancy laws to restrict the mobility of free Blacks, indigenous peoples in California, and Mexican and Chinese immigrants, among others. Some were quite general in the net they cast—the California Supreme Court held in 1854 that the state vagrancy law's phrasing of "black people" meant any non-Caucasian person[45]—but other laws made it clear whom they targeted: newly emancipated Black people. Southern states passed the so-called Black Codes after Reconstruction to arrest and then sell Black people into a convict lease system that journalist Douglas Blackmon describes as "slavery by another name."[46] Unique among laws, vagrancy functioned, writes ethnic studies scholar Hsuan Hsu, "like race . . . [it] criminalized and degraded persons for what they were, not for illegal actions performed."[47] This meant that the laws were flexible, broad, and vague enough to allow police to dispense street justice to anyone they perceived as idle, loafing, or loitering. It was not until the 1960s and 1970s that these laws finally began to be repealed or struck down, but their legacy remains today.

Digital capitalism both continues and also scrambles the equation that links the lack of work with vagrancy, because it turns things that aren't typically considered work, such as browsing idly or simply carrying a phone around as you move through a city, into tracking data and then capital. So, two things happen. First, the charge of vagrancy remains, if only marginally less explicitly, as a tool of coercion and punishment, particularly for persons of color: the two Black men arrested for waiting in a Philadelphia Starbucks in 2018, for example, gave rise to a new phrase: "waiting while black." The Starbucks incident reminds us of the structural racism that criminalizes even the act of waiting and doing nothing in an environment *designed for waiting*. At the same time, doing nothing has also become a source of profit that can be captured and even controlled by digital platforms. Thus it is harder to attend to the myriad ways that digital capitalism polices the act of doing nothing, because physical mobility or skin color or even the idea of law are at a remove. A compilation of what anti-vagrancy looks like today would necessarily need to be more imaginative; it might include, for example, prepaid debit cards primarily issued to the unbanked that charge exorbitant inactivity fees, and the Internet platforms that similarly market to the unbanked (PayPal, for example, now charges an inactivity fee of £9/€10/C$20).

But the best way to find it may be through the idea of circulation. Vagrancy is a way of controlling movement, and the movement online is, materially speaking, the circulation of images. And digital platforms refuse to let bodies of color sit still or remain quiet. In a 2017 survey of the most popular animated GIFs from around the world, the *New York Times* found that a meme of Oprah gesticulating wildly as she proclaims "YOU GET A CAR!" has become the most common way for US Twitter users to express happiness online, while Italian Twitter users express sadness most commonly through a shot of Tanisha Thomas crying on the American reality show *Bad Girls Club*, and Mexican Twitter users express humor through a shot of two fans celebrating a Golden State Warriors basketball championship.[48] All three images show overly emotional Black women—a pattern that researchers found over and over in other reaction GIFs. This pattern, cultural critic Lauren Michele Jackson argues, is characteristic of what she calls "digital blackface,"[49] where the Black body becomes an animated minstrel for white viewers—even when its expressivity isn't intended. As Jackson puts it, even "when we [Black people] do nothing, we're doing something."[50] The endless production of Black bodies as spectacle—whether "in trauma, in death, [or] in memes"—is fed by algorithms that recirculate the most viral content and thereby turn their bodies into puppet-like automatons that can be consumed as entertainment.

A meme captures and keeps a body moving to serve a wider audience; the meme's constant, looped motion works similarly to the motion of non-player characters and NPC-like characters in games, who are also tasked with manufacturing the sense of an audience, of other people around human players or agents. Distinguished by their lethargic, noninteractive, or disinterested actions, NPCs feel different or "canned" compared with the vitality that their "human" counterparts evince. Both memes and NPCs, in other words, form a complex affective infrastructure devoted to making subjects feel engaged in interaction with others online—that is, what digital media constructs as "live." And because affect is always relational, this infrastructure allows for the capacity to feel affected by others, to feel included in a generic sense of relation with other humans, to be included in the "social" of social media. These forms of "surrogate humanity" make the fragmentary publics in digital culture feel coherent, if not unitary, as the sound of canned laughter once did for television publics. But they also obscure how much of the idea of human vitality and animacy—of liveness—rests on the backs of surrogates deemed inanimate.

Liveness is, fundamentally, a temporal construct. It can look like a certain rhythm to interactivity, which marks the robot as the one who navigates a website too quickly or marks the "primitive" as someone who is too slow or passive to be part of digital capitalism. It can look like a set of temporal structures that allow some bodies to be normal and others to be marked as animated. The looped GIF or the looped NPC recirculates and overexposes bodies of color and thereby makes other (white) bodies seem unremarkable. And finally, liveness may simply be the networked rhythms that ask us to move and be moved in response to software prompts and nudges. While liveness maintains considerable purchase in our imagination of what technology should be and embody, the "deadvoice" in *Canned Laughter* suggests an alternate direction, where this world of vitality is desaturated and leaden. If the canned nature of laughter here is akin to a moment of hesitation ("lol") inside a moment of nonstop expressivity, such lethargic deferrals function not as ways of resisting the temporal impositions of digital capitalism but rather as ways of mediating this temporality differently.

This is because a "lol" is what Anne-Lise François would describe as a "recessive action," something that operates perpendicular to narrative time; a recessive action doesn't unravel narrative but typically operates in the middle of a narrative. As moments that don't develop the story, recessive actions function on the surface, working to create a different temporal structure. François focuses her study on moments of unfulfilled expectations, on actions not accomplished, and on things done with no consequence. For Okón's artwork, the point of momentary confusion in laughter that I have described as "lol" is also temporal: it is a moment of hesitation, even though the individual's laugh proceeds as expected, and even though the infrastructure of production continues to assemble into a larger, choral whole. Normally, we expect laughter to work like this: you feel something, so you express it as laughter, or, alternately, you see something funny, so you laugh, too. In this piece, the moment of laughing out loud precedes the actual feeling of laughter, when or if it comes. In this way, it builds an alternate temporality, creating space for deferring the burden of expressing oneself "authentically" (see chapter 5) or of speaking up in general.

Thus *Canned Laughter* shows us that the special privilege of being live or human is not particularly special after all. Canned laughter can make one laugh for real as much as live performance; Okón's choral performance is the performance of staleness, after all, of prerecorded emotion. To recall Wynter's argument, our idea of "the human" is overrepresented by white men in the Global North, and so to describe the human as a unique carrier of affect is also to exclude persons of color from that definition. The canned qualities of this artwork invert this equation and put the human on lesser footing than the manufactured, object-like qualities of the robotic. After all, why should the highest aspiration for digital technology be to be more vital and more human (and why should that be our aspiration, too)? The lethargy of this piece exhausts the oppressively live environment in which colored bodies are animated against their will. While lethargy describes the state of being forced to constantly move and be animated, lethargy also acts by gradually spreading out, flattening or detuning oneself from this regimen—allowing oneself to grow bored of it, to be reticent, to underperform; allowing, perhaps, lives that are vagrant or wayward to be lived nonetheless.[51]

On another level, lethargy can simply be a practice of being critically "idle"—which is to say, acknowledging one's inability to perform the hoped-for move of critique. A critic is typically driven to produce claims, to argue, to reveal, and, ultimately, to say what their objects mean. But the "recessive actions" that François asks us to pay attention to are not (intentionally) hidden, even if they go unnoticed; instead, she likens them to an open secret, which has nothing more to be revealed, but instead can be known without necessarily impelling us to take action.[52] Take "lol" again: everyone knows that a "lol" in a text message may be a somewhat disingenuous acronym, but it doesn't require much explication, because it is simply present, like a cough, in its thereness. Recessive actions instead allow us to dwell within the ambivalent middle distance which a viewer, or some of the workers, may occupy. We often miss the quieter and more reticent forms of being within and adjacent to those political moments because they are hidden in plain sight; after all, it is perhaps the ease by which *Risas enlatadas* appears as an allegory for standing up against global corporations that makes it harder to see the current of underperformance that runs underneath.[53]

But this problem is doubled when put in the context of racial capitalism. Kevin Quashie has written that because the burden of expressiveness in American culture falls disproportionately on Black people, this elides our ability to see other forms of being, such as inwardness or quietude, in the same gestures. Who, after all, is allowed to be quiet, and whose silence is otherwise seen as a threat? Black literature, in Quashie's example, is reduced to forever expressing something essential about race, resisting white domination, or otherwise offering forms of social and political commentary. But that mandatory performance of publicness eclipses other, more reticent forms of being. A gesture of defiance is clearly visible in the photos of the Greensboro sit-ins we considered in chapter 1, or in the iconic image of John Carlos and Tommie Smith raising their gloved fists at the 1968 Olympics in Mexico City to give the Black Power salute. But what we miss in calling it "defiance," Quashie says, is the intimacy of their bodies, their "graceful, lithe surrender in posture."[54] Caught in internal contemplation and thought, they embody a complementary mode to resistance: "Resistance, yes, but other capacities too. Like quiet."[55]

By foregrounding lethargy, I have similarly tried to find the other capacities at play, sometimes layered under the more obvious message or sometimes seemingly too obvious to mention. In *Canned Laughter*, I've suggested thinking not only about the larger message but also, or even instead, about the messy moments of emotional hesitation, or complicity, where the laughter produced for export may be consumed by its producers and viewers. In microwork, a lethargic approach to criticism could be to take seriously the aspirations (even if thwarted) that the workers have through their work, as well as the "open secret" that they often find something rewarding about their work, even as they are overwhelmingly aware of the depressing structural inequalities that have led them to seek such precarious work. Critics have tended to discount these experiences by claiming that microworkers must have the wool pulled over their eyes; one scholar of digital development writes, for example: "if one takes [positive experiences] at face value it provides limited impetus for reform of the digital gig economy . . . Such impetus can only be derived if one supra-interprets the evidence by attaching to the respondents a state akin to false consciousness in which they are unaware of the true nature of their labour."[56] For liberals in the North to claim to know better than their subjects what they are thinking is, sadly, one more way that coloniality works today.

Think of lethargy as deferring, rather than resolving, an ambivalent situation. We write "lol" because we don't have anything else to say in the moment, because we are required to respond, but aren't sure how to feel. But that state of deferral is valuable, too. When an artwork doesn't produce the right reactions, and leaves us interpretively idle, with "dead time," as it were, it causes the subject to decouple not just from the specific artwork but from the expectation of needing to be engaged at all: one begins to do the laundry list[57] or begins to notice one's body cramp up. This deferral might help us notice the mediation and even the racialization of subjects within a digital environment. For how much someone laughs, or doesn't, is as important to discussing race and coloniality in digital culture as its visual counterparts; the environment is something that can exhaust or enliven subjects unequally. If lethargy doesn't rise to the level of resistance or critique, it gets us one step closer to acknowledging the ordinary weight of each forced interaction, like a column of atmosphere that weighs on our shoulders, but that we have nonetheless become accustomed to endure.

SERVERS AND SERVANTS

When user interactions go wrong, our voice assistants have a plan. Alexa, Amazon's voice assistant, used to respond to a user saying, "You are a bitch," with "Thank you," but now it has a "disengage mode" that deflects such language with a "I'm not going to respond to that." Topics Apple deems politically "sensitive," such as feminism, similarly trigger a series of steps with Siri: disengage, deflect, and, finally, if pressed, quote from Wikipedia. Google's Duplex service, an AI voice that calls restaurants on your behalf for reservations, offers affective signals of fallibility and delay ("hmm," "um," "mmm-hmm") when trying to process data it doesn't understand. It reasons that what makes us sound more human is not always the most accurate response but rather phatic vocalizations. Much of the voice assistant industry's research is to turn such off-script moments back into scripted ones and, in that process, translate certain norms about emotional composure for the service industry into code.[1]

While voice assistants are just now learning how to handle difficult customers, computers have long incorporated ideas of grace and etiquette in their design, because they are rooted in ideas about server comportment.

Studying servants from the baroque period in Europe to now, Markus Krajewski argues that the computer server is descended from the human server. One key moment of his genealogy is the Victorian period, where technical devices, such as stoves and electric lights, increasingly began to replace humans.[2] Some of these fin-de-siècle inventions retain traces of their older function in their names, such as the dumbwaiter or the Lazy Susan (advertised in 1917 as "An impossibly low wage for a good servant and the cleverest waitress in the world").[3] Yet human servers had long been media (Latin for "middle"), since they typically act as go-betweens, occupying "a place in the middle, between masters and things."[4] This effect of mediation is particularly noticeable when they serve as agents of communication, as in the valets and butlers and majordomos of Krajewski's study: they relay, collect, distribute, and filter information to and from their masters.

Krajewski's butlers are part of a longer history of humans turned into media objects. In 1878, in a pivotal decade that saw women typists rapidly entering the office as an irreplaceable if invisible circuit between men and their documents,[5] Emma Nutt began working as the first woman telephone operator. Her success opened the doors to a profession that was almost entirely women-dominated by the end of the following decade. Elocution and etiquette lessons, accent neutrality, and hiring policies that largely excluded Jewish and African American women created the image of a telephone operator as a model of white femininity—"the voice with a smile," as AT&T's advertisements proclaimed.[6] Here, the quasi-secretarial work of a woman operator seemed, at the time, biologically determined: "The dulcet tones of the feminine voice seem to exercise a soothing and calming effect upon the masculine mind . . . thereby avoiding unnecessary friction."[7]

While automation (and strikebreaking) gradually eliminated the position of the switchboard operator, women continued to work as communication specialists, particularly during World War II. "Signal girls" handled communications operations overseas, as they had in previous military operations, but they were now joined by "Wrens" (members of the Women's Royal Naval Services) who adapted Hollerith punch-card accounting machines into codebreaking devices at Bletchley Park; the only African American brigade of the Women's Army Corps (WACs), who sorted a massive backlog of mail in Birmingham, England; and the WACs who were engaged in mathematical calculations, such as ballistics computations, for the first general-purpose electronic digital computer, ENIAC. Writes historian Nathan Ensmenger: "the telephone switchboard-like appearance of the ENIAC programming cable-and-plug panes reinforced the notion that . . . programming was more handicraft than science, more feminine than masculine."[8] After the war, women often leveraged their previous experience as stenographers, typists, and accountants to work their way into programming systems, but when they got there, they were expected to perform what Ensmenger describes as the "'softer,' more social . . . aspects of computer work";[9] they translated the supposedly intellectual requests of their clients into computer circuits or code.

Thus, when computer scientists drafted specifications for client-server computing, they embedded these ideals of grace into the code; as at least one white paper admits, they were thinking about their secretaries.[10] They dealt with the need to handle unpredictable input on the Internet by designing ways of handling things gracefully, for protocol is essentially etiquette: it handles interactions between strangers. For an example of this etiquette, consider the curiously old-fashioned "HELO" that electronic mail clients send to mail servers behind the scenes, in plain text (the greeting has now been revised, in a newer specification, to "EHLO"). And there is the "graceful degradation" that TCP/IP, the Internet's main communications protocol, specifies: always accept unknown input, even when it behaves badly, and when overwhelmed with traffic, try to make sure that the quality of service degrades gradually, rather than failing all at once.

With these ideals in place, the server processes that run in the background of our digital lives step seamlessly into the role of human servers, making the first ambient media not the cloud but the maid or butler. As

Krajewski points out, the postman turns into a digital mailer-daemon; the butler turns into a keyserver; the majordomo turns into the mailing-list server named Majordomo; the domain name service that translates a domain name, such as youtube.com, into an IP address, such as 142.250.73.238, was originally named Jeeves. What human and computer servers have in common, Krajewski claims, is that as an agent of media, the server must move "gracefully between presence and absence, between subjecthood and objecthood" when summoned.[11] While it is common for media scholars to claim that the server is invisible, this is too simplistic; the server must be constantly available for commands and thus can enter privileged spaces as long as it keeps moving.[12] Consumers pay for passivity: for domestic laborers, whether human or digital, to be unobtrusive, to be in the background, yet also to be available, to await and receive commands. "Those who wait" are on "permanent standby mode," to adopt computer lingo;[13] their job is to be interruptible at any time. The server is trapped inside the space of circulation, marking time until awakened by its master. Unable to either fully become a subject or disassociate from this paradigm entirely, unable to ever switch off entirely, the server is *lethargic*.

Thus technical interaction with media today has much more in common with interactions with waitstaff, customer service operators, cashiers at the fast-food counter, and other forms of service work, than with computer code. Just as computers were once women calculators who had to follow "fixed rules . . . [and had] no authority to deviate from them," in the words of the codebreaker and mathematician Alan Turing,[14] now, all sorts of persons occupy the role of "computers." They are trained to follow a rulebook and respond in a fixed way—sometimes literally being given scripts—and increasingly also trained to engage the customer in a form of generic emotional labor. In this way, the digital supply chain makes service workers into a technology. Consider what happens with a customer service agent for a bank on the telephone: even if the words they speak are not rigidly scripted (an increasingly rare exception), the agents are likely interpreting results on their screens from an opaque algorithm's decision for, say, a mortgage or a loan. Though human, they are simply expressive interfaces for another part of the digital supply chain. Alternately restricted and enabled by protocols on what they are allowed to say and how to communicate with others, they offer a very different picture of interaction than the creative vision of the "maker" that the industry commonly depicts itself as enabling. Even technologies such as Duplex are less breakthroughs in AI than reflections of the fact that society itself has become increasingly scripted.

Is technology turning us into automatons? No, but interactions between people are increasingly mediated through technological filters, protocols, and interfaces, to the extent that these technologies subtly promote a cybernetic world view. Writes historian of science Peter Galison: "The cybernetic philosophy was premised on the opacity of the Other. . . . We are truly, in this view of the world, like black boxes with inputs and outputs and no access to our or anyone else's inner life."[15] In this philosophy, each agent is a sealed "black box"—it doesn't matter whether human or animal or robot on the inside—and communicates with the next node "by the exchange of orders or commands."[16] As mathematician Norbert Wiener described in his 1948 *Cybernetics, or Control and Communication in the Animal and the Machine*, each actor is a monad, and the other is opaque: it is a "quasi-solipsistic" world.[17]

Depressing news, perhaps. But my concern is not that this philosophy is dehumanizing; instead, it is that it reinforces existing societal inequalities. Making service workers interchangeable drives down wages and turns them into a disposable workforce. It also makes the work itself increasingly repetitive, particularly when service workers are primarily interacting, in ever more scripted fashion, with literal automatons. This "black box" model allows those on the other end of the supply chain, primarily white-collar workers, to have time for unscripted or spontaneous moments of liveness, while the black boxes instead inhabit the time of waiting—that is, the time of lethargy.

Consider Duplex again: by using robots to outsource conversations with service workers, Duplex creates robot-like work conditions for them, their days solely devoted to waiting for the equivalent of voicemail phone trees and the output of other computer servers. Further, it assumes that user interactions should be made transparent and racially unmarked:[18] Duplex's public demonstration uses a Japanese-accented restauranteur as the case study for "difficulty," for what is hard to understand and process. As it converts her words into data, Duplex replaces her accented speech with that of the operating system's "accent-neutral" voice assistant. By making her business interoperable with other black boxes, such as Google Maps or Calendar, she is whitewashed by the digital supply chain. This whitewashing occurs throughout the voice assistant industry, because whiteness is coded as universal: one Black software developer acquiesces to develop an AI voice that sounds like a white woman, rather than a Black man, because he knows consumers will demand a product that is racially "neutral."[19]

It might seem that there is nothing redeeming about a moment when human interaction is increasingly mediated by the cybernetic model of the black box, with, again, "inputs and outputs and no access to our or anyone else's inner life." But rather than simply dismissing the cybernetic world view for its antisociality, neoliberal atomization, or flatness, it's worth taking a closer look from inside it, if for no other reason than to understand the complicated shifts it forecasts for society. As this chapter argues, a black box's devaluation of interiority is a provocation for us to think about liveness, privacy, and humanness—as well as its corollary, relationality—differently. What might it mean to think about technology not just as something to be pried open and revealed but as a model of black-boxing for humans to emulate? What new forms of relation might arise out of opacity, rather than "genuine" connection, that is, the revealing of one's inner life to another? And what role does gender or consent play in this admittedly cringe-inducing picture of violence delivered to humans turned into infrastructural "stuff you can kick"?[20]

To investigate these questions, I turn to the Australian feature-length film *Sleeping Beauty* (Julia Leigh, 2011), which tells the story of Lucy, a young student who juggles a full slate of server-like tasks, from waitress to clerical worker. She eventually engages in a peculiar kind of sex work (suggested by the title) in which she disrobes, takes a sleeping drug, gets into bed, and sleeps until the evening is over. She is both consenting subject and enduring object, and her lethargic, black box–like affect throughout the film, even when she is awake, will help us understand the complicated implications of agreeing to transform oneself into a server.

BLACK BOXES, EMPTY INSIDE

The first thing that one notices about *Sleeping Beauty* is how muted its mood is. Played by Emily Browning, Lucy evinces little outward emotion and converses largely in the ritual thank-yous and goodbyes of polite etiquette. The contrast between her flattened performance and slight stature with the opulent settings of her employers—lavish mansions filled with genteel quiet, apart from their elderly masters who liberally exercise their right to speak—is emphasized by the camera's theatrical framing and extended takes. (Though not quite what critics would classify as "slow cinema," the film includes a number of shots that go on for ninety seconds or more.) This means that our attention has the space and time to be drawn elsewhere: to small, even furtive gestures of acknowledgment and sociality between servants; to the presence (or absence) of breath and the other fine gradations that separate the living from the dead; to the vibrant world that flows around Lucy—students horsing around, businessmen chattering, bright red berries against concrete—as if she were an empty space inside it.

The other striking thing about the film is that despite an abundance of nude bodies, it is only nominally about sex; it is really about work, of which sex work is just one part. Even when sex does occur, the emphasis is surely on the "job" part of a blowjob or handjob. The film opens by taking the viewer through a series of vignettes from Lucy's gigs: first, a paid research trial, where a researcher inflates a balloon inside her throat that makes her gag ("You're doing a great job," he says), then a shift clearing tables at restaurant, then casual sex with a businessman whom she meets at an upscale bar, presumably for money, then a clerical position photo-copying endless reams of paper. The film interweaves these vignettes with the daily humiliations of being broke: Lucy's wealthy roommates demand rent money from her; she sneaks in behind a paying customer in the Sydney metro; there's a barely noticeable moment where she checks the payphone for spare change before she makes a call.

When she does make that call, in response to an advertisement for a high-end lingerie waitress, the film begins its second act. Lucy meets and finds herself increasingly employed by the procuress Clara, an older, immaculately dressed woman who personifies composure. Clara offers Lucy the titular work of taking a drug that renders her unconscious while clients peruse her naked body (albeit without penetration, Clara adds). These "chamber scenes" of Lucy sleeping, as the filmmaker calls them, are disturbing; they make uncomfortable and voyeuristic the viewer's desire to see what happens to Lucy. For her part, Lucy's own desire to see what the men do to her while she is asleep frames the film's final act. It ends with footage from a surveillance camera she buys and covertly installs, which shows her next to a man who ends his life by ingesting several cups of the same sleeping potion she takes.

The illegibility of Lucy's character—even, the surveillance camera sug-gests, to herself—is redoubled by the fact that the film offers little in the way of psychological motivation for her actions. Indeed, her flattened character is a common complaint of critics reviewing *Sleeping Beauty*. Writing for *Entertainment Weekly*, for example, Lisa Schwarzbaum gripes: "But Leigh doesn't have much use for Lucy's mind either: We never know what Lucy thinks about her odd job, or what she'd rather be doing."[21] She seems to act in arbitrary ways, such as having sex with men in bars, without the traditional character buildup, no doubt exacerbated by the fact that she is asleep for most of the chamber scenes. And her expression is delinked from narrative action; to borrow from Lauren Berlant, describ-ing an aesthetic of underperformance and recessive action, "The nothing might mask an event, or not."[22] The only motivation for Lucy's actions is the generic fact or open secret that she's struggling to get money.

Disconnected from other people and mostly asocial, with the exception of one friend, Birdmann, who dies midway through the film, her relationships are largely transactional. The depleted world of atomized actors that she exists in is perhaps best described by the game theory of John von Neumann, in which the "opponent acts according to certain universal maximization principles but where the thought process that eventuates in any given move is hidden from us."[23] Her psychological flatness suggests that she is simply playing a version of von Neumann's game: she is a black box out to maximize her own standing, within the constraints of the conditions forced upon her as a member of the working poor.

The freelance economy saturates everything; it even works its way into what might once have been thought of as a private life. She rehearses lines for serving her clients during one such "private" moment with Birdmann: "And how are you, miss?" "Oh very well, very well. And you, sir?" Some moments come off as ironic at first—but then we realize that, as film scholar Jennifer Fay puts it, neoliberal sleep means that we dream about work, as in her example of a Starbucks barista who dreams about muffins at the coffee shop.[24] Lucy is not just going through the motions, in the metaphorical sense that disconnects motion from intent; she is both reenacting and rehearsing the actual movements of work. Work spills over not just to leisure, as we might expect, but into moments where subjectivity becomes submerged beneath sleep. Lucy falls asleep near the copier during her office job; sleep becomes work, and work becomes like sleep. What do we dream about? Not flying, not princes, not even ourselves, this film seems to say, but rather the next job.

To be surrounded by capitalism—and to be unable to draw sharper lines between life and work—is what makes it hard to find resistance in the film. Even when Lucy is awake, she seems to be treading water. As the critic Malcolm Harris writes, referring to the "flexible" or precarious laborer that Lucy exemplifies, "Lucy shows she can endure, at least for now, but can she do more? Can the flexible resist?"[25] Yet resistance may well be just another script written by someone who is not Lucy to position her body in a more desirable state—"activity." The film is a view from the

servant class, and a critic's desire to reinscribe a more meaningful political tale onto a tale of work and survival is a desire to turn endurance into
a faster-paced event. For resistance evinces a desire for narrative change,
and endurance is too *lethargic* to look like resistance. The two terms are
imbricated, but endurance isn't as future-oriented as resistance; it looks
to the present, in its potential endlessness, and is inwardly directed.[26]

To endure, Lucy does not shatter any norms—indeed, she subsists
entirely within norms. Any act of seeming willfulness almost always
results in her eventual compliance: whether she argues about cleaning
or mocks a perverted job requirement for her lip shade to match the color
of her labia, as Harris writes, she "must ultimately follow the script in
both cases."[27] But the film takes pleasure in tracing the belabored path by
which she has to follow what is predetermined. In a single-shot scene in a
bar's velvet banquette, for example, Lucy meets two boorish businessmen
who inform her that one of them is going to have sex with her. Unexpectedly, she agrees, and coolly suggests they flip to decide who will do so.
The businessmen are visibly startled by the speed of her reply, and, after a
pause, scramble to toss a coin. When the victor exults over his luck, Lucy's
response is to delay: "Yes, my prince. But did I say when; did I say tonight,
this year, next year . . ." and when another coin flip establishes tonight,
her response is, "Now—or in five hours?" (figure 4.1). Her delay is not a
Scheherazade-esque ploy to get out of the situation that she finds herself
within—she must ultimately follow her "prince's" lead—but a miniature
game of alternately speeding up and then dragging out the process of
what everyone knows will happen.

If Lucy turns compliance into a game, it is one in which the rules are
made strange. We see this in the design of the banquette shot, which
avoids the more typical shot-reverse shot structure of two people con-
versing to create something odder. The camera pans between the two
halves of the banquette, the two men on the left and Lucy and her friend
the other, delaying the viewer's ability to see the addressee's immediate
reaction; each person takes a turn speaking, much as the client and server
communicate in turns in digital protocols, allowing the camera to explore

Figure 4.1
"Yes, my prince. But did I say when?" (The "prince" is on the left, mostly off-screen.) *Sleeping Beauty* (Screen Australia, 2011).

the distance and asynchrony within the coupling. Time runs differently here, even though, in light of her previous actions, we all suspect what's going to happen; this shot is what François, in her writing on the recessive actions that occur within narrative, would describe as orthogonal to narrative time. And an accumulation of these strange moments makes Lucy's compliance simultaneously too fast and too slow for our comfort; it suggests that Lucy is minimally coupled to the temporal strictures of those around her. Responding to Clara's suggestion to invest in her future with her money—"Pay off a student loan. Or save for a home deposit," two classic tropes of the neoliberal "good life"—Lucy instead takes her first earnings from the job and sets the cash on fire. Clara's future orientation contrasts with the present-oriented way in which Lucy attempts to tread water, without hope of a future.

I have watched the bar scene again and again, and each time she goes off with the boor. Where I hope for resistance, the film stages compliance, or what the filmmaker terms "radical passivity."[28] Even though her acquiescence there is easily explainable—a payment is presumably involved—more disconcerting is Lucy's acquiescence to other humiliations in the film, such as being tripped by a man as she tries to serve him liquor. This acquiescence leads film critic Genevieve Yue to describe Lucy as "a body that is in some sense a shell, and a subjectivity that is to some degree concealed when awake and totally hidden when drugged."[29] To be sure, the film's chamber scenes make the surface image of Lucy's body available not just to her clients but to the audience members as well, but there is no access to any sort of "within" with that vision. Nor are there more than one or two scenes where characters discuss motivation.

That we are largely unable to peer inside Lucy's character upsets one of the traditional roles of cinema as a medium for understanding a subject's interiority, as when we talk about a camera that's able to get under a character's skin, capture fleeting emotions in the close-up, or show what is normally imperceptible. Cinema, film scholar Akira Lippit argues, is like psychoanalysis in its ability to see (or penetrate) a body's hidden depths or, here, to bring its dream states to the surface.[30] Indeed, it is perhaps more accurate to say that rather than cinema as a tool for *understanding* depth, cinema, in conjunction with other modernist technologies such as photography and psychoanalysis, contributed to and cemented the very idea of interiority through which we continue to understand subjects today.

By the nineteenth century, the idea of a hidden depth within a person's image had become a consumer ideal. While oil portraits had been status symbols of the wealthy or the distinguished, cheaper photographic portrait studios allowed members of the middle class to also partake in this bourgeois mode of self-presentation. This led to something of a consensus in taste: a photo that was portrait-worthy captured a face's "fleeting expression," a melancholy or ambiguous look that offered a glimpse into a person's inner depth, even their soul. Analogous to today's how-to guides on how to achieve a good "Instagram face," studios provided an elaborate array of props and stage directions, such as looking sideways above the camera, to achieve this reveal of one's character. The idea of the fleeting expression in portraiture was, in historian of photography Alan Trachtenberg's words, the "analogue to the bourgeois idea of the discrete, autonomous, self-governing individual";[31] it allowed the viewer to sense what lay underneath even as it reinforced the privacy of the interior.

In turn, a portrait's idealization of the individual shaped the legal idea of privacy. In their seminal law review article "The Right to Privacy" (1890), Samuel Warren and Louis Brandeis discuss an 1888 case in which a woman buys a picture from a photo studio, only to find that her photographer has reprinted her face on a Christmas card he is selling. The plaintiff acknowledges that reusing a face captured surreptitiously would be legal, which leads Warren and Brandeis to argue for a different right than had been previously articulated in law.[32] That one's interior could be revealed or violated by a camera, however, both assumed a white subject and also propped up the structures of whiteness. American studies scholar Eden Osucha points out that it is a white woman's injury that is at stake in Warren and Brandeis's article, because portraits were "explicitly individuating forms of picture making . . . that affirmed whites' supposed natural endowment with capacities for 'self-elaboration' and . . . interiority."[33] At the same time, however, photographs, chronophotographs, and cinema profiled and turned the criminal, the diseased, and racialized populations into generic types: witness the eugenicist Francis Galton's use of photographic composites to "prove" that Jews were a distinct race, or anthropologist Félix-Louis Regnault's early films on the racial differences of the Wolof, Fulani, and Diola peoples.[34] Persons of color could be a commodity, or an anthropological type; they were generic, not individual.

By then, that division between an interior and exterior self had become the defining test for humanness. Post-Enlightenment thought, Neda Atanasoski and Kalindi Vora explain, distinguished between civilized bodies who registered and were changed by their environment and those whom Darwin described as "primitives" who lacked the ability to process their emotions and instead displayed them transparently on the surface.[35]

Accordingly, they conclude, Eurocentric culture has continued to privilege interiority (intent) over exteriority (display). The image of racialized subjects continues to be generic, readily interchangeable with each other, and their emotions located largely on the surface, as in the Aunt Jemima logo, discontinued only in 2020, or the memes considered in the previous chapter. And this suggests how the idea of individuality remains tethered to the idea of whiteness. Even though the logic of white injury that undergirds privacy law—often enough "the white woman's delicacy, virginity, and *intrinsic* vulnerability to wounding by the disputed media forms" (Osucha)[36]—has become marginally less explicit over time, it remains stubbornly fixed in place. danah boyd, for example, points out that Facebook's selling point in 2006–2007 was the sense of safety it offered to its (then mostly white) users; Facebook seemed to be a refuge from the scammers and sexual predators supposedly lurking in what many of boyd's interview subjects identified as the "digital ghetto" of its rival platform Myspace.[37] Nowadays, a worry about protecting the images of white women from sexting and revenge porn continues to inform the "private yet digitally constructed interiority" (Chun and Friedland) that virtual platforms conjure out of the shared spaces of the cloud.[38]

Sleeping Beauty is perverse, then, because it faithfully creates a narrative of interiority and white injury even as it is critical of it. The camera frames Lucy's slight figure in a way that highlights the innocence suggested by the film's title; the lighting practically makes her pale skin glow with whiteness; there is endless talk of discretion for the clients, setting up shots that both express the idea of privacy and allude to its potential breach. But soon enough these tropes are deflated in rapid succession: Lucy is not at all a virgin; the sex does not seem to make her vulnerable but is merely transactional; she is both objectified and yet salvages a few advantages from that position. "Poor Lucy," one might say. And yet the film skewers the viewer who might acquiesce to describing the vagina of a white woman as "a temple," as Clara puts it to Lucy. Isn't our desire to look for her hidden depths itself a legacy of cinema's obsession with knowing and depicting women's interiority on-screen—which can be

Figure 4.2
Opening shot: probing the interior. *Sleeping Beauty* (Screen Australia, 2011).

its own form of misogyny?[39] And why do we expect that she embodies some sort of distinctive character or essence, instead of a generic type—a "sleeping beauty"?[40]

A camera's ability to transgress the surface only buttresses that division between interior and exterior self that supposedly defines the human. When *Sleeping Beauty* stages the failure of cinema's ability to peer inside its characters, however, it indicates that the model of interior/exterior needs revision. The film plays with the idea of penetration throughout: in the opening shot, when an assistant in a clinical trial lowers a balloon inside Lucy's throat (figure 4.2), or when, as an office temp, Lucy absent-mindedly lowers a hole punch onto a plate, penetrating nothing. For her part, Lucy rejects the idea of a hidden depth: for example, as she enters her employment contract, she responds to Clara's guarantee that she will not be penetrated by flattening the metaphor offered to her: "My vagina is not a temple," she says. And indeed, it is not; it is just a vagina.

By the end of the film, it's clear that it has rejected a topographical model of the subject, questioning the idea that Lucy has a distinctive interiority that has been wounded—and, more fundamentally, that interiority is what makes us agentive and thus most human. We might instead return to the language of digital technology to describe Lucy's skin or shell as a kind of "black box" that bars access to, for example, an algorithm behind the glass screen and makes us indifferent as to whether it is a human or nonhuman that animates the box's input and output. This black-box metaphor is visible in the madam instructing clients told not to leave marks, even as they bruise and even stub a cigarette butt onto Lucy. It's an idea that also circulates in sci-fi television shows such as *Westworld*—where clients are allowed ultimate access to the servers' bodies, while each night a team of plastic surgeons repair their surfaces (the plot revolves around both humans and robots trying to reverse engineer the algorithm that is driving them)—or *Dollhouse*, which revolves around wiping each server's memories after each job for a client.

So, too, the plot of the film's third act, where Lucy is attempting to reverse engineer how she got a cigarette burn on her ear, why she has a limp the day after a sleep job—what, in short, happens in the middle of the night. Simultaneously, *Sleeping Beauty* asks what models replace interiority and self-knowledge, and how that registers visually. Here, the film seems to anticipate the quantified self movement, which may seem like a way of "knowing oneself," but in fact understands the self *through* its output: in other words, the self cannot be accessed through psychology or dreams or the metaphysical, but rather is another sort of function that accepts inputs and produces outputs—"you are how you act." As with the exhaustion that split Geissler into two out-of-joint versions of herself, Lucy is out of sync with—even estranged from—her own body.

Take, for example, the film's final image (figure 4.3). Shot through the static gaze of a surveillance camera she surreptitiously installs, her sleeping body is encountered by Lucy for the first time; it is a moment of self-surveillance. Her camera footage shows that both she and the elderly client are resting in bed, her face up, him face down; even though he has

Figure 4.3
Final shot: surveillance and vigilance. *Sleeping Beauty* (Screen Australia, 2011).

chosen to take a lethal dose of the sleeping drug, the distinction between the two bodies in the image is almost unnoticeable. (And, to risk pointing out the obvious, the man's corpse is played by a living actor attempting to breathe and move as little as possible, further complicating the distinction between living and dead.) Though the film sets us up to expect some sort of revelation about the body here, it ends without any sort of insight; we simply see the surveillance footage fade to black. Yue reads this as evidence of an "almost pathological" illegibility, but as I have been suggesting, perhaps it is less a pathology of disconnectedness or the failure of relationality than a proposal about the value of proximity. The final shot is footage from a surveillance camera, but that word, "surveillance," refers not to its commonplace meaning as a violation of privacy than its older meaning as vigil, a watch over a dying person. An old custom: one keeps watch through the night not because it thwarts death but because it offers the presence of another.

Bodily presence (even if an absent sort) is what the dying client requests of Lucy, and what her hidden camera does is simply to keep watch rather than reveal anything new about his character or their relationship. In the end, the film shifts its point of view from its characters to the automatic camera itself, from subjective camera to the vision of an object: the conceit is that we peer through nonhuman eyes, through the fixed distance of machine vision. The surveillance camera impassively keeps watch, even past the end of the film's narrative, as presumably it continues recording, infinitely. And that impassive look might well be more evenhanded than the film's other kinds of looking, which are saturated with misogynist violence. For the film seems to suggest that relationality does not first start with discerning the Other's hidden intentions or even knowing the Other's ontological state—whether human, robot, or something else; whether living or dead. It starts instead with the lethargic, memoryless state of input and output and protocol. As media theorist Alexander Galloway asks, "Is this [black box] just a new kind of nihilism? Not at all, it is the purest form of love."[41]

Or if not love, proximity. One common misconception about cybernetics is that it is dematerialized, but in this film, we are shown, over and over, the materiality of the body—whether it is the weight of Lucy's body as it is carried by her clients, its sluggishness when she overdoses, or its ability to endure pain—a materiality that is carried over into the film's durational experience for its viewers. Feminist art critic Heather Davis has argued that using a sex toy is perhaps one of the most embodied experiences one can have, but it is also the experience of interacting with a black box; the sex toy itself doesn't raise the question of ontology.[42] Put another way, the body of a black box need not be specifically human to function as a body or mass. Lethargy, too, can be erotic, or relational, or not, even if it is not intersubjective.

Thus the non-reciprocal nature of sleep in the film's final image hints that the bare act of being next to another is itself a novel form of relationality. For that scene refers viewers to a moment earlier in the film, when Lucy encounters her friend Birdmann, who is unconscious and about to die from an overdose. She does not intervene or call for medical help, but instead unbuttons her shirt and falls asleep with him. We see something of a lethargic or zero degree of politics here, where we have the heaviness of sleeping next to someone who can't feel anything; where affect is received if not circulated; where the nonreciprocity of the gesture involves blocked communication but nevertheless is different from solitude; where we might describe the action of Lucy as listening even though nothing is being spoken.

The two-shot of Lucy sleeping next to her dying friend is perhaps an extreme example, but it references the film's many examples of interpassivity, which I gloss here as an oxymoron: the interaction between "passive" objects that supposedly lack the capacity for action.[43] *Sleeping Beauty* often shows servants, the underclass, or other proxies of the ruling class clustered together, preparing for their work as waitstaff or sex objects. These scenes are not "lively" in any conventional sense but hint at the limited but still significant spaces in which domestic workers can be together, even as they remain in the background. While a more straightforward film might focus on the personality of each worker, thus giving them a sort of voice, *Sleeping Beauty* goes another way. It explores brief moments of sociality, for example, a long but silent car ride where a servant drives another servant, Lucy; where a servant ushers Lucy to the madam, Clara; when six working girls inspect each other or idly thumb through their phones before serving food.[44] The film is mostly silent, and shows how sex or enjoyment or liveliness is largely interpassive; everything is delegated to others, even the grooming and training of sex workers. They relate to each other not as fleshy characters with individual personalities but as servants at a formal distance to each other, and yet there is still a solidarity, however attenuated, with each other. They offer each other tea; they ask "You okay?"; they ping each other using whatever small talk is allowed by the protocol. In one of the film's most moving scenes, Lucy watches and then silently brushes a spot of drool off a woman sleeping on

Figure 4.4
Asleep on the train: Lucy watches over a stranger. *Sleeping Beauty* (Screen Australia, 2011).

a train (figure 4.4). Solidarity is created in these spaces not out of any sort of nuanced understanding of the other's inner life, but simply because people find themselves in the same place.

Counterintuitively, lethargy's coolness may better support these conditions of mutuality than active encounters between subjects, which excludes those who are unable or unwilling to speak. Here, at least, the word "passivity" may simply be a slur for radical acceptance.

REGISTERS OF INANIMACY

In the two scenes of Lucy lying next to dead or dying men, the film asks: At what threshold does an unconscious human become an object? To use linguistic anthropologist Mel Y. Chen's words, its depiction of "horizontal relations between humans, other animals, and other objects" challenges our assumptions about the implicit hierarchies within language itself.[45] When disentangling ambiguous phrases where it is unclear who is acting and who is being acted upon, language speakers parse them, Chen explains, according to what they call an "animacy hierarchy," in which language orders subjects by their capacity to be affected: men over rocks, for example.[46] Historically, however, this hierarchy divided people into categories, too, putting "rational" subjects at the top, followed by those with the capacity to feel: "women, animals, racialized men, disabled people, and incorporeals such as devils and demons."[47] But the film's two-shot of Lucy and the dead man flattens that hierarchy, making the visual equivalent of a sentence like "the girl the dead man sleeps," where it is unclear who is the subject and who is the object. By building scenes composed of objects relating to objects, not just subjects acting upon objects, the film begins to reorder traditional hierarchies about which agentive being has priority, even which being has more "life."

One place where the film reorders animacy is during the chamber scenes, when Lucy is asleep. Consider one of these scenes, where a client maneuvers her off and then back on the bed. At first the act of lifting her inert body is an ambiguous gesture that offers a hint of tenderness—a profile of a woman lifted in a man's arms, recalling the "sleeping beauty" of conventional romance (figure 4.5a). But as he sucks in his breath and continues to assess her, the word that comes to mind is plastic: both sculpture and motile, with a sheen of the manufactured. When he flings her indelicately back onto the bed, her body slips off and flops to the ground (figure 4.5b), and she remains cut off by the frame: the camera is just a little slow to tilt down, as if it, too, were growing weary. There she forms another recognizable figure of a nude reclined in repose—an echo, perhaps, of Giorgione's *Sleeping Venus* (1510)—before she deforms, once more, in his arms and he sets her back on the bed. As his exertion becomes visible and his breath labored, the scene fades to black, the motion of cinema metaphorically coming to a standstill. In a reverse of the well-known scene from Chris Marker's *La jetée* (1962) where photographic stills come to life, this is the moment when the medium itself

Figure 4.5
The penultimate chamber scene. *Sleeping Beauty* (Screen Australia, 2011).

becomes most inanimate, even deathlike. It's one of many disturbing scenes in the film that bear a distinct imprint of necrophilia.

As I watch, I am struck by the awkwardness of her being carried: for a moment, I see that tactic of civil rights protesters who let their bodies go limp when confronted with arrest by the police; they offer no active resistance, but instead force the police to face the burden of a body as a brute material object. But despite what I might imagine, the film again stages compliance. Her doll-like body can be deformed and repositioned by the client; it is weighty but also malleable, something that can be scuffed and marked, like furniture. It's no coincidence that moving Browning within the chamber scene is like moving a piece of furniture or a prop on a stage set and then accidentally dropping it. Precious, but also cheap, she resembles a human made out of plastic. To invoke Yue's earlier description of Lucy, we have "only a body that is in some sense a shell," and what Yue suggests here is Lucy's emptiness, not just as a character but also as hollow, plastic individual—in other words, a doll.

To describe Lucy as a doll surely risks reiterating the tropes around misogyny and objectification that feminist scholars have productively interrogated. Yet we should also be careful not to see agency only in assertiveness, for it is the queerness of her lethargic body—and, more generally, the queerness of being passive—that complicates a more straightforward attempt to find female agency in Lucy.[48] Instead, sex dolls offer an allegory, disability scholar Eunjung Kim writes, of how to move away from exclusionary definitions of the human as able-bodied. Rebutting philosopher Peter Singer's claim that it is "weird" to care for an individual who cannot show signs of consciousness, Kim argues that plenty of men do care for dolls—and dolls, like persons with disabilities, can inspire an "ethics of passivity." This ethics recognizes, rather than disavows, the "aspect of *object being* that enables a subject to become an object to be acted upon, and to actualize the other within the self."[49]

These aspects are beautifully depicted in the film, which stages an oscillation between Lucy's ontological status as human subject and as object. It reminds us that the categories of person and object, the former presumably having interiority and the latter having none, form a false binary. When awake, the film often asks Browning to perform Lucy in a deadface or deadvoice that seems devoid of expression. When we see Browning act asleep, recalling the quasi-pornographic Victorian custom of *pose plastique*, Lucy becomes statue-like and interacts as much with the bed or its sumptuous furnishings as with the male client manipulating her weight. That relationship between the brocade that wraps her or the chinoiserie teacup that delivers her sleep drug and the body, that suspension between thingness and humanness, constructs a synthetic being that, in Anne Anlin Cheng's words, "brings into view an alternate form of life . . . the inorganic animating the heart of the modern organic subject."[50] The film suggests that we might start with the human not as animate being but as an object that can be acted upon. This makes humanness the capacity to accept input whatever that input is, that is, the capacity to endure. By using this model—one receives, but doesn't necessarily respond—it further suggests that interaction and intersubjectivity are desired neither by the client nor by Lucy. (Sex, almost all the clients admit, is out of the question, as they are impotent.)

Lucy's seeming wish to avoid social ties comes off not as antisocial behavior but as a way of seeking a dissociative job, of seeking the self-forgetting oblivion of lethargy. Even as the social structures Lucy is caught within are coded with class exclusion and humiliation, they also offer a measure of anonymity: as long as she greets each new request with the response dictated by protocol, she is largely left alone, ignored. Each time Lucy acquiesces to a request with the lethargic response of "Okay, whatever," it reflects and produces a brief moment of emotional immobility or stuckness. It offers an allegory for today's digital capitalism: each moment of inanimation explores an alternate world to ours, where every subject and object that can strap on a sensor is in the process of being made live, connected, and animate.

Sleep is a particularly apt field for a recessive film to explore because it becomes a refuge from the imperative to interact, to smile for others, or to perform seduction, not to mention earning money through affective labor. It's perhaps this sensibility that critic Jonathan Crary captures when he describes sleep as the "only remaining barrier . . . that capitalism cannot eliminate."[51] Crary is referring to the ways in which sleep is difficult to subsume into the temporal rhythms of a 24/7, always-on capitalism. And though it is possible to disagree with Crary's literal meaning—many of my students swear by sleep-optimizing apps, while researchers have developed "dream-hacking" tools to promote memory and creativity—the direction that I take from his quote is that sleep is unassimilable because the subject can neither take control in sleep nor can they actively communicate in an economy that is built on interaction and expression.

To perform animacy, after all—to perform agency, interest, reciprocity, sociality, comportment—is unrelenting work, and to be lethargic can be a relief from that burden. Consider the film's penultimate shot, which represents a surprising feint by the director. When Clara resuscitates Lucy after what appears to be an accidental overdose of the sleeping drug, Lucy turns to gaze at the dead man next to her, and then begins screaming uncontrollably, in an extraordinarily rare display of outward emotion. While Clara clearly assumes that Lucy is upset because of the discovery she has been sleeping next to a dead man—an assumption that the audience probably also holds—the film's script indicates that belief is a misreading. It is instead the resuscitation that devastates her: "Lucy gives a pathetic cough . . . comes alive. Bafflement gives way to horror and dismay."[52] Our misreading is a result of a desire to parse out the levels of animacy here, which in turn makes the intimacy between the two bodies on the bed a scandal. Put another way, the scene is disturbing not so much for its necrophilia but because the film shows the man as an "object being"—and it shows this with dignity, even care. The same animacy hierarchy penalizes Lucy for not wanting to be a subject, that is, for wanting to be unconscious. Seen from Lucy's perspective, however, the drug is a welcome way of stepping off, as cultural theorist Matthew Fuller puts it in his book on sleep, the "treadmill of [her] subjectivity,"[53] not to mention to escape the neoliberal economy's imperative to "care" for herself or to get a life.

Sleep is the neutral meeting ground between subjecthood and object-hood, a state that shows how the two ends of the spectrum can shade into each other. The film's exploration of different registers of inanimacy suggest that "object being" can manifest in many ways, even inside those who presumably wield agency. Some of Lucy's clients seem to envy or want to imitate her doll-like or even corpse-like quality, as when the old man in the last scene overdoses on the same sleeping drug, or when the man in the penultimate chamber scene described above is visibly winded by the effort of moving Lucy, and pauses, immobile and bent over, just like her. Fatigued, the two are temporarily mirrors of each other, despite their distance (figure 4.5b). In other words, it is not that Lucy is inanimate and the others are animate; the film provocatively suggests that everybody is both inanimate and animate.

Extrapolating from films such as Jonathan Glazer's *Under the Skin* (2013), about an alien who masquerades as a human, but is hollow on the inside, literary critic Marta Figlerowicz argues: "The discovery that the lifelessness of inanimate objects is not a condition by which we are unnaturally oppressed, but rather a state that we can no longer separate from our notion of ourselves, arrives in them as a perverse relief: since we have never been properly alive, we can stop worrying about life as a quality we might lose."[54] In our context, lethargy embraces the potential of being an "object being" as one that relieves a subject of the burden of having to perform aliveness, individuality, and interactivity—all "human" attributes that are actually gendered, ableist, or racially coded. This embrace also shifts the conversation away from a narrative of redemption—of Lucy, for example, as a fallen woman who might invest in her life and pay off her student loans—because that narrative is built on reiterating and (re-) ascending the rungs of a biased animacy hierarchy. By leveling out or reordering this hierarchy, we may discover that, as Figlerowicz suggests, the boundary between human bodies and objects may be less a boundary than a continuum, possibly even that we may "exist more robustly as dig-itized pixels than living beings supposedly pursuing self-knowledge."[55] If we really want to move past the centrality of the human, we need to do so not just by making objects more animate or more vital but, reciprocally, by allowing humans to inhabit other registers of the hierarchy.[56] To take the inanimate seriously is not simply to invert the hierarchy but to examine a pessimistic strain within our social context, where bodies can be plastic, and where it is sometimes hard to tell where we are on the scale of animacy, even to ourselves.

OPACITY

Concealing one's inner life and presenting an unreadable or opaque exterior is an increasingly popular tactic to oppose a culture of mass surveillance. Adam Harvey's *CV Dazzle* project (2010–), for example, used an angular pattern of makeup that rendered a face invisible to computer vision algorithms that recognize faces based on expected features. More generally, there is a growing interest in what historian of technology Hanna Rose Shell calls a "camouflage consciousness," a growing interest in camouflaging our "true" identity from others.[57] The idea driving these tactics is a specific topography of the self, in which privacy is a cloak or shell over the individual's single interior, a piece of clothing covering a naked body, or a camouflaged exterior (known as "dazzle") that wraps around a World War II ship. Artists and activists who call for ways of camouflaging, obfuscating, or disguising one's online trail are new variations on this topography (we'll revisit this idea in the next chapter).

Seen from this perspective, Lucy's deadpan expression might be a way of concealing her true feelings from the screen. Yet the irony is that in an economic system that positions us as black boxes, our selves are already opaque to others. If Lucy is an analogue of a plastic sex doll, then, to invoke Davis's research on sex toys, "There is no mystery, no identity, no ontology, only a carefully crafted seduction . . . there is no secret to reveal."[58] Lucy already has privacy, in the cybernetic vision of the world; careful to respond to inputs with the correct output, she already has a shell and a script with which to deflect inquiries ("Very well, sir, very well"). What she needs is not privacy, at least not of this sort, and not more self-knowledge, but another model entirely.

By allowing us to explore what it means to inhabit a black box, *Sleeping Beauty* helps us revise our definition of the human: which parts of us are affected and which parts are left cold, and, in turn, which parts are animate and which are inanimate. Which parts are cultural scripts that are lodged within us? Which parts affect me, and which parts affect others on my behalf? The parts of us that I am talking about may not even reside within one's body, or any body at all; they may reside in data or in the blend of metadata that always involves others, that is leaky and not self-contained. Lucy may be a limit case, but when she turns a camera on herself sleeping, she is genuinely unsure about the extent of her consciousness and how much she can outsource it to the camera. She is like the female alien in Glazer's film *Under the Skin*, who hesitantly shines a lamp on her skin and her genitalia after an attempt at copulation, unsure what the illumination is supposed to tell her, unsure about the idea of biological sex. There is otherness throughout our being, and this is true for anyone who has had only partial access to subjectivity or reproducibility—a condition that perhaps the rest of the world is beginning to discover.

If the word "opacity" has become a rallying cry for new political strategies,[59] then let it not be simply an opacity that makes us unreadable to others on the surface, but an opacity that runs throughout. Arguing for a "right to opacity," Édouard Glissant writes: "it does not disturb me to accept that there are places where my identity is obscure to me, and the fact that it amazes me does not mean I relinquish it."[60] One is not unknowable to others while being available to oneself. One is unknowable to others because one is first unknowable to oneself. To remain in that state, in a condition of unknowing, is also to remain in amazement.[61]

"LET US KNOW SEEING PATTERNS IN CHAOS EXPLAINS"

On a summer day in 2018, I find myself marveling that a Georgian quad-rangle in London that formerly housed the Navy Office now has artists' studios inside. I pick my way through the sunbathers perched near the elaborate fountains in the courtyard, hoping to meet the writer and artist Erica Scourti, who managed to wangle space here from the arts organiza-tion. Scourti's works initially caught my eye because they didn't push back against digital algorithms, as many of her contemporaries were doing; instead, they invited algorithms in as coauthors. Scourti is trained as a moving image artist, and while moving image art normally refers to the images moving on a film or video, in the early 2010s, she increasingly began to see the circulation of images and text through digital media platforms as its own kind of moving image performance.[1] As a result, she began to openly invite Google AdWords, Facebook, and other platforms to "parasitize" her identity.[2]

In a project called *Life in Adwords* (2012), for example, Scourti emailed a daily diary entry to herself and then recorded herself reading out the advertising keywords Google's platform unearthed from that entry.

If Scourti wrote "oh god" to herself, Google might highlight this as a keyword because it might resonate with ads for Catholicism or might suggest a person's religious affiliation (and, perhaps, a person's political affiliation). The resulting series of videos offered a composite of what marketers find interesting about a person: what consumer goods might be shown to them, what their mental states are, their brand loyalties, their marital status. It showed one's identity as both personal and personalizable; in short, it reflected Scourti's status as a user back to her.

What these artworks reveal, Scourti argues, is that users have a hybrid form of subjectivity that makes it impossible to decide what is "actual" human subjectivity and what is corrupted by an algorithm.[3] As this implies, social media platforms cannot be thought of as foreign to the messages they carry; they produce subjectivity and affect as much as they circulate them. While critics have often argued that marketers, mass culture, and social media platforms infiltrate and even corrupt an authentic interior,[4] Scourti instead thinks of authenticity as a false "uniqueness . . . that is actually pushed by algorithm." I asked her about this idea, and she described her solution as instead trying to offer "sincerity without authenticity."[5] Bots, she told me, are a good example of this. For her, bots don't have any hidden agenda of manipulation; they are simply executing the logic they were programmed with. Because their interior performance matches their exterior performance, they are by definition sincere.

The critic Lionel Trilling contrasted those two same terms in 1970, arguing that the idea of authenticity developed as a uniquely modern phenomenon—one that bespeaks a self that is true to itself, regardless of what others might think.[6] In contrast, sincerity is an early modern idea of a self that presents outwardly to the public what it feels on the inside— "saying what you mean" smooths the way for others to judge what you say as true. Though sincerity has a family resemblance with authenticity, today's Western society often views sincerity as more suspect, equating it with gullibility, earnestness, groupthink, quaint affectation (the "sincerely yours" that closes a letter, for example), or unironic naïveté.[7] It's worth noting resistive agency is embedded into the idea of "authenticity": one remains true to oneself (and specifically one's choices) by pushing back against convention—or as one popular saying on a T-shirt goes, "f*ck the norm and be yourself."[8] Little wonder that the countercultural moment of the 1960s served as the background context for Trilling's study and inaugurated the "empowered individualism" that would quickly take roots in Silicon Valley.[9] In his 1996 "A Declaration of the Independence of Cyberspace," for example, Internet visionary John Perry Barlow described cyberspace as a "world where anyone, anywhere may express his or her beliefs . . . without fear of being coerced into silence or conformity," cementing the idea of the user as authentic because they are suspicious about what government or society wants us to believe.[10]

While Scourti retains some of Trilling's sense of authenticity, she tweaks the definition for the digital age to stress authenticity's link to autonomy. The etymological origin of authenticity is the Greek *authentes*, "one acting on one's own authority," from *autos* "self" plus *hentes* "doer, being," suggesting that authentic users are like sovereigns: they don't rely on anyone else to tell them what to do. In contrast, a robot has no autonomy (for now!) nor is capable of self-reflection, but is nevertheless capable of projecting affect to others, sincerely, as in Scourti's Twitter bot *Empathy Deck* (2016–17), which tweets supportive messages to those in mental distress.

What would it mean for us to be sincere without authenticity—to be, perhaps, more like a bot? It would, I think, allow us to leave behind our attachment to self-authorship and to acknowledge, as Scourti writes, that "identity emerges as much from the network and infrastructure that we inhabit and are entangled with, as it does from any sense of a coherent interior essence."[11] Spreading out, lethargically, across the mass of data in the digital environment, it would leave behind a topographic model of the self as a private interior that should be protected from mass culture (perhaps by camouflage or by disguising oneself within the crowd). And by spreading out, Scourti reminds us that, in her words, "the specific and generic, singular and multiple," are always hopelessly confused—but that this might be a productive site of possibility.[12]

In some sense, Scourti is just asking us to work with what has already happened to selfhood in an age of big data. One of Netflix's biggest technical leaps was the realization that a single account often contained multiple profiles—often because various members of a family had very different tastes in movie-watching.[13] A white paper describing Netflix's recommendation algorithm explains that it generates a diverse set of recommendations for "you," but you are a plural noun: "when we say 'you,' we really mean everyone in your household . . . Even for a single person household we want to appeal to your range of interests and moods."[14] In other words, data scientists consider "you" to be plural because a user is really a unique but ever-changing collection of data,[15] automatically produced by technologies of managing populations, such as ID cards and passports, credit scoring, usernames, or the purchasing databases compiled by data analytics firms. And we users, in turn, learn to become legible subjects by adopting those technologies of management, for instance, by presenting our ID card when asked, or by authenticating ourselves with a username and password.[16] (This fundamental reorientation leads philosopher Gilles Deleuze to propose that we use the word "dividual" rather than "individual" to refer to the self, which has become divisible and combinable.[17]) Traces of the user's plural origins are immediately visible in the logic of personalization, in which algorithms recommend choices from other users who like similar things. The model of selfhood here is simultaneously personal and also based on a theory of social grouping, known as homophily, which describes the tendency for similar people to seek out each other.[18] The user may well feel individual, but the user also bears traces of the population that surrounds it.

While our multiplicity is obvious to big data algorithms behind the scenes, platforms take great pains to hide it and to address "you" as an individual, because many users perceive the presence of other people's choices—or even choices offered to a generic demographic category—as a threat to their individuality. As artist, writer, and curator Aria Dean points out, an "opposition to the desire to constitute ourselves as complex, individual subjects . . . would contravene the inheritance of 20th century identity politics [which] taught us that one of our rights is a right to representation, not only politically but personally—that we have a right to be represented as we are. But what if one says to hell with that?"[19] Dean, along with other critics, such as Lauren Michele Jackson, have pointed with dismay to the unequal circulation of bodies of color online, with Black bodies in particular turned into memes of animatedness. As we saw in chapter 3, the overcirculation of their images makes them less individuals than a generic type interchangeable with others, whether representing emotion, trauma, or simply excess. (This is a reminder that some bodies are always more "entangled with" infrastructure than others—indeed, that some bodies *are* the affective or even human infrastructure for digital networks.[20]) Yet exploring the possibilities of being generic may be an alternative to the strong forms of identity—Dean cheekily terms this "full-bodied high-res representation"—that constitute the raw materials for digital capitalism.

As my friends and I joked after the 2009 news that HP webcams can't track Black people's faces, and again after the 2019 National Institute of Standards and Technology report that facial recognition systems are 10 to 100 times more likely to misidentify Black and Asian faces than white faces, what's so great about being distinguished, tracked, and surveilled by a camera, anyway?[21] Isn't going untracked finally a small reward for being seen as inscrutable, of being one of a group that "all looks the same"?[22] Of course, it's just a joke; a misrecognized face in a security camera can be deadly, and there isn't really a way of escaping, in the end. But the point isn't that we need more accuracy. It's that making our identities as visible and represented as possible also means making ourselves legible to the carceral systems that feed on such visibility.

One result of embracing what happens when, as Édouard Glissant puts it, "one consents not to be a single being and attempts to be many beings at the same time" is that what we think of as collectivity changes as well.[23] If one sets aside the insistence on the uniqueness of the individual, collectivity would be no longer tied to an idealized form of networked publicity that would rely on the mutual exchange of individual interests[24] but rather something more lethargic. As this book has begun to suggest, there is a rich set of states in between sociality and antisociality in networked culture that has gone relatively unexplored, including timepass and being proximate with others. They are part of what art historian Kris Cohen terms the new "group forms" of today, which come out of the constant negotiation between the technically generated populations of digital capitalism and the democratic ideal of the public. Describing a project that scrolls through the random queries of people simultaneously logged onto a search engine, for example, Cohen detects an "atomized, nonreciprocal togetherness, togetherness that is only barely not the same as being totally alone"—and yet that crucial phrase "barely not the same" makes all the difference: it is still something.[25] While it is common to argue that technology produces forms of inauthentic connection that either make us isolated and "alone together" or produce echo chambers of people saying the same thing, Cohen's example hints that the new forms of relationality are already flourishing *within* the space of digital capitalism, rather than outside it.[26]

Scourti's artworks similarly show that certain intimacies manage to circulate through and across the filters and gated communities of the Internet. For instance, in an exhibition called *So like You* (2014), Scourti uploaded her old vacation photographs to Google's reverse image search. Vacation photographs often resemble each other, and this has become even more intensified in an age of mass image circulation; think of how many Instagram travel pics repeat the same gesture of a person in a mid-air leap, or posing with feet dangling off a cliff, or sitting on top of a car with a mountain backdrop.[27] But rather than comment ironically on this resemblance, Scourti decided to email others whose photographs, the algorithms had decided, were visually similar. Her correspondents traced the outlines of an otherwise-invisible group that is produced in parallel to personalization's explicit communities (on, say, social media) and implicit populations (in marketing databases). To choose strangers based on the visual similarity of their vacation photos may seem like an arbitrary way of making a connection, but it nevertheless offers a new kind of proximity that is orthogonal to one's "likes" (even if not entirely outside them): it discovers a group form that had been hiding in plain sight.[28] What's especially enjoyable about this project is that it takes the part of the algorithm that produces "likeness" literally, even sincerely; the visual likeness between two images creates a moment of felt connection. The project embraces the ways that one is connected by what Cohen calls a "penumbra of affective togetherness," connection across the long voids of the Internet.[29] Each image Scourti found is less unique and individually authored than a sign of sticky connection, suggesting the possibility that the algorithms that pen us into gated communities are also producing new forms of collectivity across those boundaries.

I also find something poignant about Scourti's use of old vacation photos to be "like you," given the different moments in time from each person that the algorithm brought into alignment. Think about the strange time skew that results from using Google Street View, where one can time travel between a photograph Google took of a certain street from 2016 to another one taken in 2021 by simply advancing a few feet on the same block. Because a lethargic subject's identity is coauthored by algorithm, the temporality of a lethargic subject is subject to similar moments of time skew, not to mention the vagaries of networked circulation. Pieces of the self might move faster, becoming a compressed, degraded, or "poor" version that circulates more readily across the back channels of the Internet,[30] but these pieces might also languish, altogether unseen and unviewed. The self races ahead of the information that is online: as I write this, I think of not just the numerous outdated photographs of myself floating around, but the uncanny way in which the text of some writing I did a decade ago has been scavenged and recirculated anew as spam. But it also lags the version of me that is online, which is recirculating and being recomputed and resold as I sleep and which I will discover only later, in limited, fragmentary forms, perhaps through an advertisement or a search result, or perhaps through a connection that someone makes later. I am not in sync with myself, but I don't think I would want to be. Just as personhood emerges out of the entanglement with network infrastructure, so, too, does the temporality of personhood. Lethargy is the unwinding of the normative bonds of time that constitute an individual, as well as the feeling that results when trying to reconcile this with the feeling of progression and coherence expected *in* the individual. There is always a part of me, in other words, that is lagging.

We can see some of these temporal disjunctions in Scourti's performance *Think You Know Me* (2015–), which takes advantage of her iPhone's predictive text feature.[31] By repeatedly and furiously tapping the screen to get the next word the algorithm suggests, Scourti generates a sort of loose script for an impromptu monologue (figure 5.1). As Scourti races to speak words that can connect the algorithmically generated prompts (as well as occasionally live input from other users on Twitter) into a disjointed narrative, the stream of words on screen continues to march on. The predictive text comes from several sources: Scourti's phone has trained its dictionary on her emails and input, and Scourti has also saved a few prewritten phrases, such as "Robot servants are coming to make your life easy," as texting shortcuts. The result is an emulsion of live voice, screen text, and what is pre-programmed.

Artist and critic Nathan Jones has pointed out that the viewer of this performance is caught in a constant struggle between the visual text, which appears quickly on Scourti's iPhone screen, and Scourti's slower attempts to keep up as she reads that text. The autocorrect algorithm is faster than she can think or enunciate, making her behavior reminiscent of harried office workers who continually apologize for the "delay in replying," to invoke a phrase Scourti uses in the performance; at several moments the pace amps up the weird glibness of certain phrases ("soon soon as possible so that everything will soon dissolve"), as if Scourti were a fast-talking salesperson. But here it's important to note that the performance hinges on more than just the algorithm's speed. While Scourti's earliest performance of *Think You Know Me*, at the 2015 transmediale festival in Berlin, relies on the impossible speed of the algorithm—which paints a futurist picture of ever-accelerating machines—a later version at the Haus der elektronischen Künste in Basel, Switzerland (2015), slows

Figure 5.1
Screenshot of iPhone text from Erica Scourti, *Think You Know Me*, performance at transmediale festival, Haus der Kulturen der Welt, Berlin, January 28, 2015. Courtesy of the artist.

down and speeds up in response to the language on-screen, providing a different view of time, as if this image of the future were already beginning to decelerate.[32] By training her eyes on the screen, the viewer can partially imagine the temporality of predictive text: the viewer is always faster than the performer; and yet the viewer quickly realizes that this "real-time" speed is decoupled from the phenomenological world. As the viewer toggles between the visual and the auditory, she embodies the author's inability keep up, not just as an index of a depressive subject's "inability to keep pace with life" as lived by other people (Jones),[33] but as an index of her lethargic inability to keep up with herself.

While predictive text technologies evolved to simplify text input on cellphones, they are also *social* algorithms that make predictions about what an English- (or Spanish- or Arabic-) speaking user might want to say next, personalized to the needs and lexicon of a particular user. Thus the text that appears on-screen refracts, through a distorted lens, not just Scourti's typing (and her personal habits of speech) but also the generic phrases of a mass/multitude represented within her phone's dictionary: "a while to reply," "time and money," "do a pic," "in my opinion." These personalized dictionaries not only power the autocomplete function on a smartphone/computer, but have also evolved into Google's new Smart Composition feature, which helps a user write email by suggesting phrases that might complete a sentence, or even entire responses that it has observed a user write previously ("Thanks so much!"). They work well not just because these mechanisms are getting quite good, but also, as we saw in the last chapter, because mobile correspondence is often scriptable and routinized ("get back in touch," "love you too," "heyyyyyyy"), and, arguably, because we live in a textual environment so full of automatically generated (or translated) writing that perhaps our sense of textual coherency has shifted to accommodate such former novelties: we don't flinch when occasionally the ghostly tics from what appears to be another user's personality appear in "our" email.

We see a clue to this emulsion of individual and social even in the title of the performance, "Think You Know Me," which curiously lacks a subject. Is it "I" (Scourti) who thinks "you know me," or "you" who thinks "you know me," or even a third subject (an algorithm, presumably), who thinks "[it] know[s] me"? This is a sly reference to the fact that it's unclear who is even authoring the text: another moment where the text's "authenticity" is impossible to determine. In the performance, Scourti barely keeps afloat on this endless stream of language recycled from her prior text input history—a previous "me"—and from other peoples' input histories. Autocorrect is both automatic (or machinic) correction and also a continual process of auto (or self-) correction. In this way, language is unquestionably an instrument of socialization and subjectivity, and it is sometimes hostile in this performance, like a timekeeper inside a factory. (Indeed, a pop-up notification on her phone that says "Timer done"—presumably to signal that her performance is ending—infiltrates one performance, resulting in her intoning "times times time's UP" even as she attempts to ignore it.) But the found language of autocorrect/autosuggest also offers a way of slipping out of the straitjacket of subject and object: here it is unclear who (or what) is acting on what, who wields agency and who is on the receiving end of it.

These traces of non-agentive behavior are clearly visible in the Basel performance. Often enough, it is as dry as an office memo, as when Scourti notices that her phone's autosuggest function keeps offering her the same sets of phrase choices, and begins to deliberately cycle through those phrases: for instance, between "cheapest price" and "best price," or "for delay in replying." But even this banality can turn into something profound: "you can't help me either it's the best it's the best way to the best way to look at the best way of okay," she says, and its off-kilter humor comes out of the seeming contradiction of "the best way of okay," which contains the melancholic sense that the speaker is seeking the best way of feeling okay even as that way is just okay at best.

Jones has suggested that Scourti's work is an example of glitch because it heralds a new kind of subjectivity or language, but I see something else going on. While Scourti's texts may resemble the aesthetics of the glitch, or the poetry culled from spam emails or other internet detritus that is known as "flarf," both of those genres rely on the extraordinariness of an event or of language, that is, where a system for producing language has gone awry. Glitching and the flarf movement have marked themselves as self-consciously avant-garde practices that take advantage of deliberately ugly language or aesthetic extremes; in flarf poems one might find a mash-up of Hitler and unicorns, or the eponymous figures of Sharon Mesmer's "Squid versus Assclown" (2006). In contrast, much of *Think You Know Me*'s impact comes from the ordinariness of language, where algorithmic language and SMS abbreviations have become diffused into everyday speech ("OMG"), and where machine translation or predictive text or applications of natural language processing are ubiquitous.

In this post-Internet world, it is thus traditional poetic techniques that highlight the interstices in that ordinariness. Poets use enjambment in verse, where a phrase or sentence spills over the end of the line into the next without any punctuation, to simultaneously suggest the separation of the lines on either side of the line break and also the joining of those lines into one thought. Something similar happens in Scourti's performance despite the lack of line breaks on the screen.[34] Because Scourti rarely pauses to start a new paragraph, generally preferring to type and speak as if it is all one sentence, the lack of lexical demarcation causes the viewer to find multiple meanings within the phrase. Is it "delay in replying can't feel statistically / if you were born" or "can't feel / statistically if you were born" or even "replying can't feel"? Here's where mundane phrases that are built by concatenation can peel away from their scripted meaning, such as apologizing for an actual email delay in replying turned into an apology for not feeling.

Scourti's technique produces both ambiguities in meaning and also the sense that phrases join with and abut each other in strange ways, as if an imprint of all the ways in which language has potentially been (re)assembled by others. This logic of concatenation is helped by the fact that there is no going backward for Scourti's spoken words. Any recall of a previous phrase, however designed, is partly by happenstance; Scourti must wait to repeat a phrase until an algorithm offers it to her again. Words (and speakers) mostly accumulate here, one thing after another, and the result is a jumble between traditional notions of a lyric speaker and other voices spoken by anonymous users. Writing about a formally similar artwork— Omer Fast's *CNN Concatenated* (2002)—film scholar Erica Levin argues: "Concatenation creates a figure that conjures the unrepresentability of the mass in negative form."[35] In other words, these stutters in language index the moments when the genre of personal communication lapses back into the wider population from which its language was drawn, as if the populations created by algorithms were being unstructured or declassified into their informal components.

✻

Though the shift to digital media has made the idea of a single mass audience obsolete, there is, nevertheless, an affective pull or "relief of being in a mass."[36] This is because a mass, theorist Tiziana Terranova argues, is by definition asocial (rather than antisocial); it is a group of people who simply figure as intensities of feeling. The mass has not "learned who we are and where we belong . . . [or] assum[ed] a role that is defined for us by another subject"; network culture, in contrast, is a site where everything has become socialized.[37] (Even the idea of individuality, as I have been suggesting, is a role taught to us through the process of socialization.) While aware that a mass has changed irrevocably in an age of networks, Scourti's artwork, I am arguing, summons the intensities of an imagined mass out of the populations that surround her.

Mass culture has historically had a vexed relationship with avant-garde art (as well as authenticity); it is often stereotyped as feminine, because of its subjects' supposed tendency toward passivity, consumption, and distraction.[38] Scourti decides not to make her language cohere into a single "authentic" voice, but instead performs distractedness by reading off the incoming (and sometimes random) text on her phone's screen, such as pop-up notifications and tweets from others. And the language she uses is often borrowed from and mediated through the mass culture of consumption; in the Basel performance, the phrase "robot servants are going to make your life easy" recalls the way washing machines and vacuum cleaners were marketed in the 1950s and 1960s to women, and, more broadly, an imagination of female domesticity that continues to inhabit the gendered voice assistants of Alexa, Siri, and Cortana. The speaker of the performance is in some sense a robot servant herself, who must follow the text of her machine-generated "autobiography" as a chore—just like Alexa and her colleagues, who have autobiographies written for them by their corporate owners.

And if, in this stereotype, a female artist is supposedly too close to her subject or too embodied to stand at a distance from it objectively, Scourti turns this excess of proximity into a virtue: not only are "her" words literally embedded within a field of mass culture, but she also takes on the frequently gendered emotions of "awkwardness, shame, boredom, and anxiety," as one curator summarizes it.[39] As a case study, consider *Woman Nature Alone* (2011), a long series of videos that mimic stock video footage, of the sort that an advertiser might purchase to fill out a scene: "A girl in tears. Emotions of a young woman. Close up," or "Sad woman sitting by the edge of a cliff."[40] These stock videos are art objects, but they also have a second, parallel life: since Scourti houses the videos on YouTube, they appear to any YouTube user searching for, say, sad women.

Performed with sincerity but not authenticity, they allow Scourti to emote as the system asks, without necessarily being "authentically" sad (which is to say, having an interior reason for being sad). The figure of sadness here is just "emotions of a young woman," rather than a specific person, and yet the sadness is performed without parody or irony—indeed, judging from the YouTube comments, it is occasionally capable of genuinely moving viewers. In this process, the subject is transformed into "Young woman," a stock figure that can be purchased by videographers or repurposed by others.[41] In this sincerity, Scourti is also not critical in the traditional sense of the word, since the work's criticality rests within the distributive circuits and new contexts found by other users. Her stock footage is silent on what it is saying, because it is both commercial and personal, individual and *generic*.

To disconnect sincerity from authenticity is therefore also to delink feeling from a specific subject: feeling therefore becomes a free-floating attribute, an intensity rather than a specific emotion. Videos such as "A girl in tears" and "Sad woman," along with another one of her artworks, *Screen Tears* (2010), which played a tape herself crying on a television for sale in an electronics store, show how Scourti is interested in evoking the melodrama of daytime television programming—of soap operas and other "weepies" aimed at female audiences. Critics frequently describe melodrama as a low genre of culture because of its tendency to produce a sympathetic, embodied response in its viewer—on-screen tears beget off-screen tears.[42] These "weepies" are seen as evidence of how the masses may be manipulated through emotional excess, with their emotionality a sign of their unsophistication. Here, however, the melodrama of Scourti's work is denatured. If Scourti's performances are moving image works, it is not Scourti's specific tears that matter; they create an affecting but also generic icon of sadness, like an emoji, that detaches emotion from a specific user so that it can be circulated. Scourti's techniques give voice to the "penumbral" affects that seem to circulate within populations, those ghostly moments where we see the traces of someone else's way of speaking. These moments recall mass culture but are not reducible to it; in their de-differentiated status, they provide a form for lethargically occupying a position within a population.

The affects that circulate between populations lag behind the speed with which they are continually recomputed and recomposed. Yet they offer a *feel* of proximity that is orthogonal to the "liking" or homophily of populations, as in Intel researcher Melissa Gregg's metaphor of "data sweat," a metaphor born out of observing conference-goers sweat—their bodies affected by others' heat. Writes Gregg: "Sweat literalizes porosity: it seeps out at times and in contexts that we may wish it did not. . . . Sweat leaves a trace of how we pass through the world and how we are touched by it in return."[43] Rather than projecting one's supposedly authentic (and thus private) self outward, one adds to, and also picks up on, the sense of feeling that is circulating, that is "in the air."

What Scourti describes as creating "sincerity without authenticity" is an ambivalent approach to digital capitalism. Rather than withdrawing from it or resisting it, sincerity suggests that it might be possible to obey digital capitalism's injunction to speak up while slipping temporarily into something more lethargic than full subjectivity. Scourti's performances speak, but in that speaking, explore the effort and exhaustion and hesitancy of being (and maintaining) the role of a speaking subject. "I am not myself these days," the speaker of her videos might as well be saying, but this is because other voices and other selves are at play at the same time, other selves that are echoed back to us, algorithmically, and also a self that is decomposed by algorithm into attributes (race, gender, income, location). These moments of frustrated or hyperactive speech cross what would normally be the most personal of genres, the confessional, with moments where the performer temporarily reverts to a mass-like state. Just as sincerity, as critic Adam Kirsch explains, "can only be manifested in relation to other people, because it involves meaning in your heart what you say aloud,"[44] a *digital* sincerity might try to find company with others within the limited forms of connection offered by digital capitalism, for instance, by exploring what group forms are inadvertently possible at the same time that the algorithms of image searches, predictive text, and social media feeds individuate a user. By doing so, it might jog the memory of the networked user that has for too long been induced to desire a public built out of individual agency and choice, and remind it that other ways of forming the social are still possible.

FEELING PROXIMATE

Despite the flood of investigative exposés and scholarly articles on surveillance and big data's material infrastructures, users in the United States seem more enmeshed than ever in those infrastructures, as if they are responding with a collective shrug—"I know, but all the same." It's not just naïveté on the part of users; it's more accurate to think of big data as something that we notice but that we don't really remark on. For an end user still registers—if peripherally—the fact that one is being profiled. This isn't always an explicit awareness, though that can certainly happen when personalization becomes too obvious, or a selling point, in the case of TikTok. In the main, though, users respond to personalization largely as an often-inchoate, barely registered, but intense feeling of being "normal," which is to say, comfortable, included, and connected to a generic sense of social life. The affective flush of "feeling normal" helps explain why we identify with digital platforms and subject positions even when we know they have the potential to harm us.

When describing media, being normal is almost always seen as a negative, a sense that a mass audience has lost a sense of individuality and given up their capacity for aesthetic judgment. This concern is a long-standing one: "They all surrender to American tastes, they conform, they become uniform," wrote one German publisher in 1926 about the effects of the American film industry on German audiences.[45] "Normalcy" also suggests a kind of discipline: people must conduct themselves according to norms of behavior, or else. For this reason, "normal" often describes an imagined set of behaviors that one can use to blend in, such as the popular "Gray Man theory" sites, which offer advice on what to wear to disappear into a crowd when planning for a breakdown in civil society: "natural and neutral colors work best; Browns and grays. Nothing to create a memory like a T-shirt with a saying or photos. . . . Ordinary is the key word here,"[46] or the web browser plug-ins such as TrackMeNot that obfuscate your search terms by spewing out a cloud of more normal-seeming search patterns gleaned from other users online—for instance, "Living fulfilling lives without; that have been lost; Make friends with your; must push back against; Jaundiced newborns without additional."[47]

Of course, the very notion of "acting normal" depends upon exclusion. To "blend in" or to pass as normal assumes a certain kind of privilege; some subjects—persons with disabilities and persons of color—have already been marked, sometimes violently, as outside the normal.[48] This is why the artist Leo Selvaggio, a white man, modeled his "normal"-looking prosthetic facial mask—for users to present as a generic white man to a surveillance camera—on his own face (*URME*, 2014). He proposes the mask as a way for someone to "temporarily experience and consequently perform White male privilege."[49]

While these tactics are suspicious of or take advantage of perceived norms, "feeling normal" is a powerful affective state that helps us navigate the confusing world of digital capitalism. For that construct captures the sense of feeling both individual and public at the same time; it is, as media scholar F. Hollis Griffin writes, "an experience of freedom and belonging; it is both a flush of recognition and a fantasy of generality."[50] In the sense of queer theory that Griffin builds on, "normality" supplies scripts and narratives for living that makes one feel included in a generic sense of social life. In an earlier era of mass media, for example, television shows proffered the fantasy of home ownership or a heteronormative family in ways that many white audiences recognized as comfortingly close to their own, "normal" lives. Today, the affect is captured by social media and its effect of personalization, which attempts to recognize and situate you within a larger, general context of others who share your interests. Data broker Experian's Mosaic product for digital advertisers, for example, has around seventy discreetly named market segments for each country it operates in, such as America's M44 "Red, White, and Bluegrass" and S69 "Urban Survivors," or Britain's N59, "Asian Heritage," that act as proxies for race, ethnicity, family status, and income and serve to recreate the "neighborhood" each household occupies. (At one point, when an airline data leak accidentally embedded a visitor's Mosaic scores inside a hidden part of the webpage, I discovered I was in segment G24, "ambitious singles," people who in their description "carry rolled-up rubber mats to work, prepped to duck out at lunch for a yoga class." Reader: I wish!)

One might argue for any number of feelings that big data is said to produce (or does in fact produce), and almost every marketing research firm offers a slightly different one, from friendship to relief to interest. But perhaps the best evidence for normalcy as a baseline mode is that almost every marketer describes what happens when personalization goes wrong as "creepy" or "uncanny." In a well-known, cringe-inducing case from the 2000s, Target sent coupons for baby clothes to a teenage shopper that its algorithms believed to be pregnant. She was, but she hadn't told her father, who figured it out when he saw the coupons, and was outraged.[51] The idea of being in a demographic neighborhood filled with identical interests can be sufficiently claustrophobic that big data algorithms used by Netflix and other recommender systems occasionally introduce a mistake or a recommendation that they think you won't like, so that you aren't weirded out by how uncannily well they seem to know you. By inversion, big data tries to produce the "at-homeness" that the uncanny (*unheimlich*) unsettles in the original German: familiar, belonging, comfortable, of the family, intimate, even if this is effort is so often riddled with failure.[52]

For this process of personalization to work, one must first identify with one's digital avatar and, more fundamentally, with one's "authentic self"—one's "authentic" emotions (such as "likes" and "dislikes"), consumer preferences, and affinities. In return, the algorithms will return other people and consumer products that fit those affinities. Lest this be understood as a natural chain of events, keep in mind that there are many other ways of constructing social algorithms. A 2018 summer workshop in Amsterdam explored social networks, such as Mastodon, that allow for a user to register multiple identities for each community they play a part in, obviating the need for a single stable "authentic" self; alternately, a developer could connect users based on mutual indifference, rather than mutual affinity.[53] As Wendy Chun suggests, mutual indifference might produce social networks that are organized less around intensities of emotion and other types of content designed to trigger strong reactions than ones that are deintensified. The result would be social networks that are a little less like a gated community where everyone is expected to hate the same things (e.g., Fox News or CNN) and a little more like a city where it is possible to walk by another person with disinterest—a social form organized mostly around the emoji 🙂 rather than just 🖤 or 😣; a sociality that is more asocial or, to use the terms of this book, more lethargic.[54]

One idea digital scholars tend to accept, and which this chapter has attempted to dispute, is the fundamental idea that we should identify with our own "likes" and our own behavior, rather than, say, what is deemed socially appropriate to feel ("sincerity"). This conception of selfhood, Trilling and others have argued, is a relatively recent phenomenon of modernity. While Freud and other psychoanalysts laid the groundwork for models of selfhood in which desire and subjectivity are both constitutive and at odds, the sociologist Eva Illouz has shown that this model is also embedded in the framework of what she calls "emotional capitalism" that began to cement itself in the Western workplace in the half-century after the 1920s.[55] Attempting to help workers better communicate by registering and managing emotional states, these management theories were indebted to the relatively new science of psychology, and specifically the idea of recovering an "authentic self" that is hidden from oneself, but can "apprehend itself through texts, classify and quantify itself, and present and perform it publicly" through technologies of analysis (whether therapeutic or, as Illouz argues, digital self-fashioning).[56] Emotional capitalism is now one of the most noticeable features of social media platforms, which encourage us to classify our emotional reaction to a post through a limited set of responses to further capture our behavior: on a Facebook post, do you "like" it, or do you feel "love," "haha," "wow," "sad," or "angry"? In today's age of big data, a user's past behavior even serves as signals that suggest how we might act in the future, and becomes a way of finding ourselves.

Undergirding this is an ideology that says we should naturally want to be ourselves, that our preferences and emotional states are of interest, and that we should naturally follow them. In short, personalization teaches us to identify with our wants, and uses this to find other wants. Of course, it is theoretically possible to refuse this sort of identification. But because these data technologies are widely permissive—they count *everything* as "normal," as allowable—they create a universalism that promises and by rights ought to include everyone.[57] Those users who are uncomfortable with this universalism are caught in a double bind. If mass consumerism continually asked, "Don't you want to be happy?,"[58] the equivalent of refusing this system is saying no to the question "Don't you want your own wants?" or, alternately, "Don't you want to be yourself?" To say no is perverse, self-sabotaging.

I mean to evoke here feminist scholar Sara Ahmed's idea of "melancholic universalism," which she defines as "the requirement to identify with the universal that repudiates you."[59] The universal ideal of connection and authenticity should automatically include this user, even as it rejects them for not participating successfully in that ideal. Though Ahmed's words refer, for example, to the experiences of being marked as a colored body within the universality or default value of whiteness—a universality one could try to join by renouncing one's family or other cultural marker—or, in another example, to a conservative gay man attempting to become "normal" by claiming he believes as much in family as do homophobes—I believe they also apply to the digital environment. If we are positioned to identify with our wants (as re-presented to us by behavioral data) to "feel normal," the result is a proliferation of melancholic universalism throughout the digital environment: when a platform forces a trans person to identify with a deadname in the name of becoming an authentically "verified user"; when credit scoring requires someone in debt to identify with an opaque data profile and, ultimately, a credit system that excludes them; or when employees of Uber or other ride-sharing companies are required to identify with their platform (i.e., their identity as Uber drivers), even as that platform exploits them. The violence here is not simply of being forced into certain patterns of behavior or being encouraged to buy more products, but of making a certain form of "authentic" subjecthood that we call the "user" stand in for the universal. The user—and all the straightening mechanisms contained within it—is one of the most ingrained if unquestioned building blocks of digital infrastructure. And if we try to escape it, it can feel that we are repudiating ourselves.

What I have described as "feeling normal" online exerts a pull because it explains and gives meaning to the otherwise-confusing double address that makes each user simultaneously an individual and also part of a population or demographic body. Particularly because big data constantly recalculates and resynthesizes the population that a user is compared to from moment to moment—digital scholar John Cheney-Lippold gives the example of an algorithm deciding you are 60 percent African American, then, after a few more clicks, 80 percent—it must also shore up the sense that your behavior is part of a larger collective body.[60] When it succeeds, it says that no matter what you do and where you go and who you are, "you" are part of a community of people who think like you.

The sense of feeling "normal" points to the way that personalization mobilizes the affective charge of mass audiences, even though there is arguably no longer a mass audience in an age of big data, nor a single "normal" position for a user to adhere to. Some websites handle this reality by coming up with ersatz ways of offering the traces of others, so that, for example, a travel website such as Expedia will inform you that "10 other people are also looking at this hotel," even though it is almost impossible to ensure that other people are really online and looking at the same thing at the same time. That notice is, of course, a nakedly consumerist way of manufacturing artificial scarcity and competition to spur you to book the hotel, but it is also a fascinating way of manufacturing a sense of co-presence, of suggesting that you are in a temporal relationship with others.[61] It suggests that we are all in a global relationship with the rhythms of consumer demand, or news, or even online community—a rhythm of enforced sociality that crowds out the ability to be asocial or alone.

But if the travel website can create its own competitive way of making you aware of others, other ways are also possible. As Scourti demonstrates in her exploration of canned language and stock figures, we can let go of our claim of being complex, individual subjects in search of others with common interests, and still find ourselves together, however lethargically, with others online. In fact, we always already are: before one gets into a discussion of how to construct communities out of individual subjects, the first step is to recognize that the subject is already constructed out of otherness. Even when nobody else is around, their preferences and their traces still linger inside us. And those traces suggest a kind of being together that need not rise to the level of a public, because they have not yet been socialized to become one.

PORT SIGNALS

For scholars who think about the social impact of technology, one of the most interesting shifts has been the way that control increasingly works through movement, rather than by restricting movement. Users are less and less like schoolchildren disciplined to sit still than like animals in a game reserve, to use a metaphor from Gilles Deleuze, "free" to move about while tracked by an electronic collar.[1] With mobile phones, this reserve has grown to encompass everywhere with signal, and the user is not just monitored, but prodded: the more a user moves through a city with their device on, the more that their location may be correlated with their historical behavior, and in turn nudged toward another location.

A layer of physical movement animates digital networks, no matter how dematerialized they may seem. For they are in essence a series of logistical systems that locate, position, and coordinate goods, information, and bodies alongside each other, binding users and suppliers and servers from one continent to another. Amazon, for example, is not just a website or a network of server farms, but the barcoded system that coordinates its inventory and positions its 350,000 robots and 1.6 million workers in a delicate, almost balletic dance—as evidenced by a video its robot manufacturers made of the robots performing to music from *The Nutcracker*.[2] If we want to better understand these networks, we should examine not just the lightning-quick movement of an Internet packet across the globe but also the disputed movements of warehouse workers lobbying for ergonomic equipment, of ships idling and dockworkers waiting to unload their cargo—or going on strike. Withholding movement is a central tactic of organized labor, and much of the organized resistance to globalization after 1995 has come from the transport sector, such as dockworkers in Brazil, teamsters in the United States, and railway workers and truckers in France.[3] But it is harder to effect a stoppage of logistical networks today, which are increasingly able to turn stoppages to their advantage. Witness the business model of digital cloud providers, which sell "self-healing" networks and disaster recovery, or the container ship that blocked the Suez Canal in 2021, sending shipping rates spiking and contributing to the ship operator's jaw-dropping, 29-fold increase in profit six months later. Disruptions benefit today's economy of intermediaries. How, then, to proceed?

If digital systems can be thought of as choreographing a complex series of motions between human and nonhuman workers, dance choreography can offer tools for thinking through them. This chapter uses the ideas and techniques of the dancer and artist nibia pastrana santiago to examine the constraints and affordances of physical movement in a digital age. Exploring how colonial legacies are embedded in today's logistical systems, pastrana's work suggests both the impossibility of not moving, particularly for a colonial subject, and also the necessity of it. This lethargic deadlock is productive, this chapter shows, because even in the absence of a formalized political movement, it politicizes ordinary movements done (or not done) collectively if not intentionally, even as small as the act of idling together.

I'll start this by describing one scene that encapsulates for me the double articulation of performance and media infrastructure. At around 1:51 p.m. on a hot day on June 6, 2019, pastrana lay down on a sidewalk on

Figure 6.1
nibia pastrana santiago, *objetos indispuestos, inauguraciones suspendidas o finales inevitables para un casi-baile*, Whitney Museum of American Art, June 6, 2019 (author photo).

Gansevoort Street, New York, outside the Whitney Museum of American Art (figure 6.1). On this sidewalk, perfectly framing her head and feet, were a set of fluorescent orange arrows and brackets spray painted on the sidewalk marked ECS ("Empire City Subway") and "No Digging" that indicate fiber optic cables below her.[4] This action took place about an hour into what would eventually be a five-hour performance. By the end, she had draped herself with electrical cables from a speaker broadcasting a soundtrack that combined spoken words with an ambient but at times aggressive, piercing drone. She had dragged those speakers and cables so that they faced the pipes below the grating of the sidewalk, causing an eerie resonance from its underground cavity, and then positioned herself in the middle of those cables. In these actions, she positioned herself as part of the circuit between dancer, audience, and the infrastructure that makes lower Manhattan one of the densest collections of fiber-optic cable and communications equipment in the world. The final hour saw her—surprisingly—rip up the cables that surrounded the periphery of the outdoor space serving as her performance venue, cables that were taped neatly aside or smoothed away under black and yellow cable ramps. This action directed our attention to the support structures for the performance that lay there in plain sight—perhaps like the maintenance workers taking out the trash, who also hovered in the peripheries of the scene—but which few audience members had given any thought to.

Cables were not the only circuit, though, that ran through the space of pastrana's performance. Near those markings were discreetly hidden surveillance cameras, which she pointed to in the middle of the performance, and two bollards, those steel pipes awkwardly installed into the sidewalk that began to mushroom after 9/11 to provide a measure of security or crash protection from rogue cars. A few minutes after she had laid down on top of the fiber-optic junction, pastrana sat against one security bollard and rested her chin against the second's uncomfortable metal surface, creating a strange kind of canceled "rest," and a strange dialogue between the ideas of rest and security. As she "rested," a concrete stone embedded into the sidewalk announced "Whitney Museum of American Art Private Property Access is by Revocable License Only," delimiting the invisible border on the sidewalk between private and public property, and how urban space is now licensed like a app's end-user license agreement— that is, using it means consent. As a performer in the Whitney Biennial, pastrana had negotiated the right to lie down on that sidewalk beforehand, but it was a right that could be, and almost always was, revoked for most bodies of color that wanted to lie there. (As the plaque also inadvertently suggested, American Art was also private property. Access to it was similarly temporary and revocable.)

The securitized but thoroughly ordinary atmosphere was not just confined to the immediate surroundings. Clearly visible to the audience looking to the right of pastrana were ships passing through the Hudson River; as pastrana performed, the soundtrack occasionally crackled with VHF marine radio traffic between ships and the Sector New York command of the US Coast Guard, or, at one moment, audio from a Fox News segment about the heroism of the Coast Guard's first responders. As she told me during the three-week research period leading up to the performance, a billboard advertising Royal Caribbean Cruises and the warmth of the Caribbean islands hovered above one place she was practicing, another link—albeit a grotesque one, given the relationship between colony and colonizer—connecting the space where pastrana had been commissioned to perform "American Art" with her home and workplace down the Eastern Seaboard, San Juan, Puerto Rico. During this period of research, pastrana walked along fences marking the water's edge and construction zones, and laid down and performed an act of "self-eroticization" on a series of New York Police Department concrete barricades.

These maneuvers began to tease apart the infrastructures for control and communication underneath the museum and its surrounding environment. While pastrana counted at least four visible layers of fences between the museum and the water itself, each one referring to a certain type of securitized space—a museum, a pedestrian zone, a construction site, a foreign trade zone—space was not the only thing that these barriers controlled. Time was also layered, too, and alternately enabled by and also regulated by these infrastructures: the leisure of joggers next to the laser networks for high-frequency stock trading next to the idleness of persons without housing. During the performance, pastrana variously twerked with a traffic sign marked "No Standing, 8AM–Midnight, All Days"; laid under a service van marked with the logo of Temporary Walls, Inc. while pointing to the word "temporary"; and used windsocks to capture the ephemerality of the maritime atmosphere, where the passage of time is registered not just as a number on clocks but also through fluctuations in the wind speed and the temperature. As pastrana has stated, "The main material in my work is time. I believe in its strategic use to provoke risk, or boredom."[5]

For a choreographer, time might occur in sequencing, in repetition, in duration, rhythm, velocity, movement: a repetition of a movement, for example, sets up an expectation for the future. pastrana's "strategic use" of time is perhaps most clear in the five-hour duration of the performance, which incorporates the exhaustion of both the performer and the audience into its structure. The act of lying down on the concrete stairs, for example, became a very different gesture when repeated three hours later. After the baking heat, her body was sunburnt and visibly covered with sweat; the ground that at first seemed to provide some respite became a hostile environment.

pastrana's performance was more than the performer enduring something difficult, however. As performance studies scholar Sandra Ruiz has written, endurance is a permanent condition of life for the Puerto Rican subject having minimal political or economic power and thus trapped in a state of dispossession and exhaustion.[6] Endurance is how one "does time" when one has too much time and is also, simultaneously, running out of time. And as pastrana whiled away her five hours, she also drew the audience into an event of endurance. This is not to say that the audience underwent anything as demanding as pastrana; the audience was free to leave at any time. Rather, the performance was structured to expose audience members to many of the same elements of the environment as pastrana, so that they had a shared experience of the traffic, the heat, the visual and auditory experience of security. For example, at one point, pastrana went out to the edge of the road, and many of the audience members followed her out there, too, forming—largely, it seemed to me, unintentionally—a perimeter between her body and the road. When she pointed at the security cameras embedded in the wall, we in the audience, mostly paying patrons of the Whitney, became aware of our own acquiescence with its regimen of security, and whom that regimen keeps out.

And as she brought out a windsock and the audience watched the weather together with her, alternately bored and fascinated, we began to be aware of our own idleness, even the absurdity of it all: a group of people standing around underneath a "No standing" or "Do not enter" sign. pastrana's idling was contagious, as if it had made us all lethargic.

objetos indispuestos, inauguraciones suspendidas o finales inevitables para un casi-baile (indisposed objects, suspended inaugurations, or inevitable endings for an almost dance) brought to a head much of the work that pastrana had done earlier on circuits, ports, and security, largely done in and around the San Juan Bay. The bay is a good location for examining global logistical networks, because, as pastrana argued in a short publication about the choreography of ships, "the bay is protocol."[7] Protocol, Alexander Galloway writes, is a "type of controlling logic" that manages the flow of data on a logistical level, and though he is writing here of computer protocols, such as ones that govern how data flows through the Internet, this captures pastrana's sense of the term as well.[8] In comparison to the ports of New York and New Jersey, the San Juan Bay is constricted by heavy-handed measures, such as the Jones Act, which bars non-US flagged ships from carrying cargo to Puerto Rico, and causes the cost of importing basic products—food, oil, and energy—to skyrocket. But access is also a function of standards and infrastructures. In a poem titled "Datos sobre la bahía," Pastrana suggests that the bay is literally studded with technologies for documenting "weight, quantities, measurement" and surveilling and registering merchandise and location.[9] And then there is the bay itself, which is being continually dredged and deepened, and the surrounding lagoons filled. If you listen closely, you can hear the way the maritime language of flow, packet, and channel is shared with network culture: even the boats' continual "*entradas y salidas*," entrances and exits, also translate as inputs and outputs to be recorded, taxed, and controlled.

Consider how these shipping networks became the starting point for a choreographic event. In pastrana's *El Weather Bureau* (2017), once a week for four weeks, she led fellow dancers and collaborators through

Figure 6.2
nibia pastrana santiago, *El Weather Bureau: circuito coreográfico temporero* (2017),
premises of the old US Naval Reserve Officer's Beach Club, El Escambrón,
San Juan, Puerto Rico. Left to right: Juan Carlos Malavé, María de Azúa, Lydia
Platón, Gabriel Maldonado, Adriana Garriga-López, Aneek Uhuru. Photo: nibia
pastrana santiago. Courtesy of the artist.

a series of exercises on the roof of an abandoned US naval officer's club
overlooking San Juan Bay (figure 6.2). In one exercise, performers used
the flag encoding system of the International Code of Signals, the system
used by vessels to signal each other. In another, dancers were divided
into pairs and separated, and then one person would create a gesture
with their arm, something that would "signal based on the information
[they] are perceiving" through their body—about the weather, the sun, or
the wind, for example. The other would repeat it, or not, testing the idea
of what bodily signals could be transmitted across the distance. And in a

third exercise, pastrana asked her performers to echo earlier movements so that the movement would cycle across the circuit. "Is signaling a score for control? Is the body an instrument for forecasting?," pastrana has written, but if the body is indeed an instrument for a forecast, she uses it as a deliberately unreliable instrument.[10] Unlike mathematician Lewis Fry Richardson's 1922 vision of 64,000 disciplined human computers arrayed to form a massive circuit for calculating the globe's weather, pastrana's collaborators misread and misinterpret each other's signals. The idea of forecast in *El Weather Bureau* became an experiment in the idea of both anticipating and being out of synchrony with others.

There is also a specific history pastrana recalls when asking bodies to perform as signaling infrastructure. Puerto Rico was the first Latin American node in the telegraph network, although this occurred somewhat by accident; Samuel Morse often wintered in Puerto Rico with his eldest daughter Susan Walker Morse, and during one such visit in 1858, he laid a two-mile line in Arroyo from a warehouse to his son-in-law's sugar plantation. Morse hoped the system would grow to eventually connect Europe with the Southern US states, and while the Civil War derailed this plan, the United States rushed to build a telegraph network as soon as it claimed Puerto Rico back as a territory. The island's location made it a key maritime checkpoint; as historian of technology Tomás Pérez Varela explains, "Puerto Rico was the first place where a boat [from Europe heading to the Panama canal] stopped to stock up and the first place where information was obtained about its load. This information, which included data of various kinds on the products transported, was extremely valuable in commercial terms."[11]

These telegraph systems were also invaluable for their ability to provide an early warning of hurricanes in the Atlantic.[12] In the late nineteenth century, the Caribbean telegraph network, writes historian Stuart Schwartz, "now made possible the dream of the weather watchers: simultaneous observations over widely separated distances and the creation of synoptic weather maps. These visions seemed to promise predictability. States could see the utility for agriculture, maritime commerce, and war that such a promise implied."[13] However important, the predictions from these weather stations were marred by the colonial administrators' distrust of the "lazy" peasantry in the Caribbean. Schwartz recounts that the devastating Galveston hurricane of 1900—considered the deadliest storm in US history—was successfully predicted by Cuban meteorologists. But this information was discounted by the US Weather Bureau because, as with the US administrators of Puerto Rico, "they felt that the Cubans were too emotional and interpretive, and not wedded closely enough to mathematical readings: their reports could not be trusted."[14]

In the analog work of forming these "temporary circuits" in Puerto Rico and later in New York, one thinks of the cablemen on the British Empire's insular colonies who retransmitted telegraph signals to the next node on the network; one goal of deploying these cablemen, media scholar Nicole Starosielski writes, was to educate the colonized in the proper behavior and proper protocol of their colonial masters.[15] The network bound not just territory to empire or periphery to center but also bodies to the rhythms of global capital; the British began to derive profit from both the trade of commodities and the coordination of bodies across the globe. Today digital capitalism relies largely on the network effects of the latter—the assembly of a critical mass of users—for its profits. The epigraph from E. M. Forster's novel *Howards End*, "only connect," has become the slogan for marketing campaigns and social media networks. Indeed, connection often substitutes for a message. Using the example of the app Yo!, which has exactly one function—sending the word Yo to someone—philosopher Robin James explains: "it doesn't matter what we say, only that we ping one another, that we establish patterns of relationships, patterns of behavior, patterns of circulation" to generate capital.[16]

But there is another, more redemptive way to read the temporary connections that pastrana's event evokes. I previously described the use of the flag encoding system of the International Code of Signals in one exercise. What's striking about the Code of Signals is how many flags are simply about the matter of communication: "I wish to communicate with you"; "I am disabled; communicate with me"; or "Stop carrying out your

intentions and watch for signals." In the performers' repetition of essentially phatic moments of sociability, I am reminded of writer Amos Oz's
anecdote of his parents savoring the thought of making a long-distance
telephone call to relatives each month. When the time comes for the
call, however, the conversation is simply about the fact that they will talk
again soon. Oz's story, media scholar John Durham Peters argues, is evidence that communication is more than the transmission of meaning;
it is a way of exchanging "tokens of presence" and existence.[7] Similarly,
El Weather Bureau is clearly not an exercise in efficiently transmitting
information—if anything, it resembles, in a playful way, the child's game
Telephone, where a circuit is formed between children and the message
gradually degrades as it is passed from ear to ear. pastrana's piece is a
way of asserting the proximity and, arguably, the connectedness of each
actor in the space. Luxury yacht and cruise ship and container ship and
pleasure craft are there, on the ocean, as well as individual dancer and
local resident. Drawing on the maritime infrastructure that supports all
of them—satellites, flag systems, buoys—as well as the medium of the
ocean itself, *El Weather Bureau* insists on the connection between ago-
nists, even between things that are indifferent to each other.

As pastrana writes elsewhere, in *maniobra*, "There have been immigrants who, when they arrive at port, are thrown from the cargo ships
into the water. . . . Everything that enters or leaves is considered merchandise."[8] This declaration is more than the obvious comment on how
commerce reduces everything into merchandise; rather, it is the precon-
dition that would allow a solidarity to develop between everything that is
considered "merchandise": the immigrants; the cargo; even, potentially,
the tourists, who are themselves merchandised by the tourism industry
(as in the industry term "self-loading cargo," meaning passengers). In
this context, it is difficult not to also think of Harney and Moten's observation that the first items of "merchandise" distributed by modern logistical
networks were the bodies of enslaved persons, extracted and removed by
colonial powers and then shipped across the Atlantic Ocean: logistical systems were originally technologies of colonialism.[9] If we know anything

about their lives, it is not through biographies or histories but through shipping manifests and insurance documents. When immigrants are thrown into the bay, as cargo that is less valuable than cargo, this history repeats itself again.

EMBRACING DEADLOCK

While scholars are still grappling with the histories of Spanish colonialism and the American military presence in Puerto Rico, there is a newer layer of history to take in account, what economists Larisa Yarovaya and Brian Lucey have described as "crypto-colonialism."[20] Borrowing a phrase once used to describe countries that are nominally sovereign but financially dependent on donor nations, they describe how the financial allure of cryptocurrency and blockchain has led the Puerto Rican government to promise investors a depopulated island base to carry out experiments for remaking the monetary system and the power grid. Even functions traditionally performed by the state, these entrepreneurs claim, should be run by "smart contracts," contracts automatically governed by digital protocols, rather than by, say, a judicial system. And, of course, a further incentive for investors is that they can buy private access to the bay: the government, one investor said, is selling "docks and an amazing view of the waterfront."[21]

Taken together, it's an experiment that involves abandoning existing public infrastructure that keeps the elderly and vulnerable alive in favor of private, market-based infrastructure that may be unavailable or unaffordable for the populations that need it most. As one capitalist told a reporter, "It's only when everything's been swept away that you can make a case for rebuilding from the ground up."[22] The result has been government policies, such as tax exemptions for entrepreneurs that are unavailable to locals (Act 20/22), that have resulted in a wave of immigration by US capitalists, who are almost entirely white men—and a corresponding migration out of Puerto Rico, as these latter-day settlers drive up rents and persuade the government to abandon its poorest constituents. The island is uniquely situated within and in the shadow of the US economic power, where the airplane to Florida by which many Puerto Ricans have commuted is nicknamed "la guagua aérea," the flying bus; American studies scholar Marc Priewe calls this transnational way of life "dwelling-in-mobility."[23] Thus crypto-colonialism offers a case where the ideology of free movement within the digital world works its corrosive effects through the mobility of bodies in the physical world: in other words, where a seemingly new "control society" interfaces with old-fashioned settler colonialism.

If crypto-colonialism and other logistical fantasies work through movement, can they be resisted or refused? pastrana entertains the idea of a work stoppage by harbor pilots as an answer, and compares her lying on top of fiber optic cables, sidewalks, and police barricades to a strike—sort of. Many contemporary Marxist geographers concerned about today's diffusion of logistical violence advocate for causing blockages; reasoning (with only partial success) that logistics relies on continuous, smooth flow, they call for creating chokepoints in roads and digitally occupying servers.[24] And we could certainly interpret the image of pastrana lying in the middle of infrastructure this way. Indeed, in one clipping from her blog that documents her research, pastrana shows dockworkers in 1952 refusing to offload perishable cargo; in another entry, she displays a photograph printed in *El Mundo* from 1969, captioned: "The photo shows the inactivity that reigned in the port of San Juan early yesterday morning. . . . For reasons unknown, [the twenty ships expected] did not arrive."[25] These historical examples inspire her to enunciate a tantalizing, if brief, idea: "the bay is always open . . . what if we closed the bay?"

But pastrana's body of work shows that the idea of inactivity or closure she is working with does not have the same literalism as traditional Marxist tactics of blockades or stoppages—indeed, they might not be tactics at all. Her work, instead, shows us a direction that is neither resistance nor refusal, but is, to reappropriate the loaded history of idleness in Puerto Rico, *lethargic*. This idea of striking, after all, is circumscribed by the political realities; in recent years organized labor in Puerto Rico has been less and less able to effect political change through strikes. The government has subjected these unions—already internally divided—to a myriad of pressures, including "fiscal emergency" laws that allow the governor to suspend union contracts, tax code changes that have drastically reduced union membership, and a law, Law 45, that seemed to recognize public unions but made striking not only illegal but a pretext for decertifying a union.[26]

The larger issue is that, as with other colonial populations, even the performance of inactivity in Puerto Rico has had a poisonous history. The Spanish colonial governors criminalized vagrancy through much of the nineteenth century: an 1838 law declared that anyone found without "income, occupation or honest way of living" would be impressed into forced labor, and a later law made it mandatory to carry a document called a *libreta*, which supplied daily proof that one was gainfully employed, and recorded one's "attitude" while working. Liberal reformers in the late nineteenth century repositioned laziness as a problem of medicine and hygiene, blaming racial mixing with Black former enslaved persons as a leading cause of indolence.[27] And almost as soon as they captured the island from the Spanish, US administrators began to frame the native population as a "lazy, easygoing, and in the main, idle people" and brainstormed ways to transform them into "active and energetic workers."[28] (As the Galveston weather-prediction example showed, this didn't make the administrators trust the subjects of US colonies any more.) Lest we think that this racist legacy is buried in the past, recall President Trump's tweets after Hurricane Maria, where he declared that "they want everything to be done for them."

As pastrana told me at one point, "Who gets to do nothing in Puerto Rico? The tourists." The Caribbean islands package laziness as a commodity for tourists, but make laziness unavailable to the actual residents. Caught in between being impelled to move and the exhaustion that results from doing so, the result is the affect that I have been calling lethargy. So the idea of inactivity as resistance is less a literal option than a seed for a performative one, as the question "What if we closed the bay?" indicates:

"closing" a natural feature is no more possible than stopping the waves. As she writes in a pamphlet about her Whitney performances, "Mi interés por la coreografía tiene ese feeling of strike, de huelga. Quiero decir: suspended inaugurations," carrying the *feeling* of striking into another set of terms, "suspended inaugurations."[29] As we discussed this idea in the spring with the sound artist Eduardo Rosario, who was experimenting with the soundtrack to pastrana's Whitney performance, he recalled a childhood full of temporary or provisional solutions that had become permanent: a can of beer that props open a window due to a broken window-slat, because a proper repair will never arrive; nobody will ever fix it. For him, everything in Puerto Rico is suspended like that temporary fix. One spends days forever in between, he told us, receiving a new promise of financial repair every four years, with another election, but of course the result never comes.[30]

These moments of suspension create intensely deflating political moments, but they also hint at a new framework for thinking about temporality and action. To describe this framework, pastrana invokes the Barbadian poet Kamau Brathwaite's idea of a "tidalectic" history in the Caribbean—a history that moves cyclically, like the tides themselves—rather than a dialectical one that moves toward a resolution. Brathwaite has described the idea through a memory of watching a woman sweeping sand from her yard, a continuous back-and-forth motion that seemed, as he looked at her, to evoke the constant travel and return of bodies between Africa and Jamaica in the middle passage: "like the movement of the ocean she's walking on, coming from one continent / continuum, touching another, and then receding," says Brathwaite, suggesting how the movement invokes the network of relationships between all the points touched by the ocean.[31]

This is more than just a textbook critique of progress and linearity. Geographer Jonathan Pugh (also quoted by pastrana) explains that tidalectic currents "work by . . . bringing about moments of impasse and states of suspension, while at the same time creating something new."[32] Giving the example of the slow evolution of the colonist's language into creole or pidgin, Pugh argues that moments of impasse and suspension—moments where nothing seem to happen—are part of a system of slow transformation. Impasse does not directly lead to or cause newness, just as boredom is not automatically a creative or revolutionary act. Rather, a deadlock, or the lethargy that is its affective signature, is a sign of an encounter or friction between multiple modes of time. One can't force it to be productive, but one can, as pastrana does, at least stage the friction.

pastrana weaves suspended action into her Whitney performance, but it is perhaps the central issue in *taller de nada* ("Workshop on Nothing"), the performance I turn to now.

LETHARGY AS A GROUP FORM

In 2015 pastrana invited several collaborators to join her in a workshop titled *taller de nada*. Her artistic collaborators—Rosario, choreographer Siriol Joyner, sculptor Elizabeth Robles, theater artist Nelson Rivera, poet Nicole Delgado, performance artist Karen Langevin—were joined by two guest speakers: anthropologist and artist Adriana Garriga-López and philosopher and artist Bernat Tort. Together, at the Museum of Contemporary Art in San Juan, they used their art practices and mediums to think, work, and do "nothing" together, for three hours a day over a ten-day period, with participants and audience members free to come and go during that period. With simply that idea as a guideline, rather than any pre-formed notions of what they would produce or how they would act, they interpreted the idea in a myriad of ways. Robles, for example, installed string sculptures that seemed to sag under their own weight, while Joyner was present for the whole time, trying to find movements that would somehow "do nothing" next to, but not in conjunction with, pastrana, who was next to her (figure 6.3). Meanwhile, as there was no physical separation between the audience and the collaborators, audience members often moved through the space that would normally designate a performance—occasionally stopping to rest, mingle, or simply observe.

Figure 6.3
Frame from video documentation for nibia pastrana santiago, *taller de nada*, Museo de Arte Contemporáneo de Puerto Rico, Santurce, Puerto Rico, May 26–June 7, 2015. Videographer: Laura Patricia RA. Courtesy of the artist.

The scene was, in a word, lethargic: participants (and the occasional audience member) sagged onto supplied pillows and disappeared into them, sprawled out onto the wooden floor, or simply slumped against the walls. But it was hard to simply hold one position for long, particularly if others were in the room doing "something," so participants would shift again, trying—and often failing—to do nothing. As I have put it, lethargy is both the inability to move as well as the necessity to move, and this restlessness gave the workshop its distinct character. A foot tapping repetitively against the floor, while the rest of a body lay supine; a person rolling listlessly left and then right on the floor; leaning back against a wall and then straightening up again; hitting the wall with a palm of the hand: many of the motions resembled the repetitive movements people make when antsy or otherwise are forced to be immobile, to wait. Thus the participants are doing nothing in the sense of *accomplishing* nothing, of "busy idleness" or acedia: their motions are the equivalent of flicking pages on a phone, or twiddling one's thumbs.

The lethargic movements within *taller de nada* were notably different from stillness—which requires a great amount of physical effort to arrest muscles and maintain position. For Joyner and pastrana, the two participants explicitly trained as dancers, these repetitions also loosened the disciplinary claim that dance makes upon the dancer's body, told to assume a certain posture, flexibility, or simply not to fidget, for example. When we first met in New York, pastrana explained these repetitions by tapping a nearby chopstick wrapper repeatedly, telling me that they were ultimately a way of canceling out the initial motion. Just as a word said over and over dissolves into meaninglessness and pure sound, a motion, done over and over, in pastrana's practice, can negate itself. It is, in other words, a way of using motion to cancel out motion, rather than simply remaining motionless.

This idea of cancellation fed into an investigation of alternate structures for action outside of "doing." At one point, pastrana asked participants in the workshop: "I want you to feel an impulse to do something and then not do it." This instruction can evoke various ideals: perhaps delayed or deferred gratification, perhaps self-denial or asceticism as a meditative cancellation of desire, or, what I think is most apt, the fact that action need not always be actualized for it to have an effect in the world. Latent, the etymological twin of lethargy. In pastrana's instruction, I am reminded of the self-canceling structure of the novel *The Princess of Clèves* (1678). In this story, the princess falls in love with a duke but is unfortunately married; when, after years of waiting, her husband finally dies, she surprisingly elects not to pursue the affair with the duke. This choice is typically read as self-denial, gendered frustration, or passivity. Yet as Anne-Lise François retells the princess's story, it is instead a "recessive action": a modest or slight action that reveals how predominant an ethos of actualization is—indeed, to the point that desire is conflated with the pursuit of desire (even if—or particularly if—it fails).[33] The Princess's non-election contrasts with the refusal of Melville's story of Bartleby, a scrivener who "prefers not to" work or move. How both protagonists are received is gendered: the Princess continues on her path as is, circumscribed in her options, while Bartleby stubbornly stakes out a different path; this is why today's scholars celebrate Bartleby for his defiance and resistance, while the Princess's abiding barely registers as an action at all. Yet in François's words, "*La Princesse de Clèves* repudiates fulfilment and assaults instrumental reason and its corollary, the hope of completion or possession."[34]

The workshop's room occasionally echoed with the sound of footsteps or idle movements, with the occasional vocalization or phatic utterance, and even with maintenance workers entering the space. Rosario also set up speakers and performed some of his sound art live. But pastrana had asked the participants not to speak to each other during the three hours; instead, she supplied them with papers, markers, and tape, and when they wanted to express something, they wrote down something and then taped the text to the walls. These short texts—which she likened to titles— were often acts of self-observation ("titling" the present state) or thoughts on the idea of nothing: for example, "Is nothing repeatable?," "Hartos del espectáculo," "Volverse nada." Visible to all who cared to look, they were statements that did not require a response, although during certain times designated for discussion, these titles often served as prompts for the participant who wrote them.

Perhaps the most striking titles were two that declared the participants were engaged in a process of "hacer nada colectivamente," or "do[ing] nothing together." While idleness is normally read as a personal or moral failure or, in the case of tourism, individual pleasure and enjoyment, doing nothing would reposition idleness inside the framework of the social. For starters, the participants explicitly interpreted the topic of doing nothing in the context of political action. At one point during the workshop's discussion, pastrana told me, the question came up about what it meant to be doing nothing in the middle of the island's economic crisis. But the question "What should we do?" and doing more generally, critic Nelson Rivera pointed out, was always framed in terms of what its colonial occupiers wanted from Puerto Ricans. Since the financial crisis of the 2000s, economists have prescribed a flexible labor market, and accordingly the Banco Popular of Puerto Rico launched a campaign in

2011, "Echar Pa'lante," which translates into something like "Move forward." This campaign urged listeners to reject "reluctance; pessimism; indifference . . . laziness." As historian Mabel Rodríguez Centeno notes, the bank commissioned a well-known salsa group, El Gran Combo, to rewrite their own song, "No Hago Más Ná," a famous "hymn to passivity" that even featured in pastrana's workshop. Now, the group turned the song about doing nothing into a hymn to work.[35] "Doing," then, cannot be separated from the context of empire, where doing is often quite literally associated with possession—whether extracting land or bodily labor or economic potential.

Doing nothing collectively would thus be a kind of politics by other means: one that is not centered on an individual refusal to act (again, as embodied in Melville's story of Bartleby) but acknowledges and negotiates the limitations of choice. For an example of this, take the 1998 plebiscite on the status of Puerto Rico, which asked voters whether it should declare independence, become a state, choose sovereign free association, or exist as a commonwealth with limited self-government. A majority of voters voted, by 50.2 percent, for "none of the above." As feminist scholar Laura Briggs reads it, the winning choice of "none of the above" is both literally accurate—voters expressed dissatisfaction with the options presented, particularly its definition of commonwealth—and also a "poetically accurate expression of an appropriate refusal by Puerto Ricans to believe that anything important could be resolved in a status plebiscite," given that the US Senate had already declared its opposition to the vote and that the United States has a history of violently suppressing previous independence movements.[36] And, as Briggs points out, "none of the above" acknowledges the reality of US economic sovereignty over Puerto Rico. For her as for anthropologist Yarimar Bonilla, who has written movingly about the Caribbean's "non-sovereign politics," "sovereignty" is its own kind of liberal fantasy of progress, agency, and independence that washes Puerto Rico from US hands, despite a shared and entangled history.[37] We might point out that the complex entanglement shows up in plebiscite as a set of bad choices, which results in neither a "yes" or "no," agreement or refusal, but rather deferring the decision. While the governor had called for settling the status urgently and permanently, the vote, conversely, called for postponement. Its citizens settled again into the process of waiting.

The idle movements of the San Juan workshop seemed to embody the myriad ways that waiting is a matter of occupying suspended time together. And here is where pastrana's workshop structure produced what might be described as a lethargic form of collectivity. In a shortened, ninety-minute version of *taller de nada* I attended in Chicago in the spring of 2019, where all audience members were participants, I observed our ten-person group form occasional moments of unintentional rhyming with our bodies and gestures. For instance, I moved to a corner, and I discovered a few minutes later that my corner had grown into a semi-circle of people. (While dancers are often trained to be attuned to the movements of others, as in the modern improv technique of "flocking," only one of our group besides pastrana identified themself a dancer, and both dancers were moving on their own.) Similarly, some new bodily postures or gestures seemed to teach other people new ways to be lazy, even though we were instructed not to explicitly respond or cooperate before the workshop began. The way a gesture or movement circulated in the room seemed akin to the involuntary contagiousness of yawning or laughter: despite one's best efforts, one finds oneself yawning in response to someone else's yawn. This form of being together does not rise to the intersubjective dialogue that makes a public sphere, but it is phatic or affectively proximate. Sociologists instead call this kind of collective behavior "herding," and it is typically used to describe the irrational, non-agentive, animal-like behavior that characterizes a crowd stampeding or a mass of naïve consumers. But in Chicago, it appeared in its benign form: "doing nothing" in a herd, we seemed less human than idling animals, but we nevertheless formed a kind of collectivity through mere proximity, rather than sociality.

This sense of proximity is different than a heart-warming story about community. Joyner, who was present during the original *taller de nada*, told me that she was initially unsure how she fit in to the space of the workshop; on its first glance, pastrana's "doing nothing" did not seem to differ from the way that, say, John Cage's *4'33"* prompted its pianist to do nothing. But that feeling for Joyner changed after Robles, the sculptor, shared a letter from a Puerto Rican political prisoner, Oscar López Rivera, during a discussion period.[38] The letter described his wish to see Robles's work in a museum someday, but as Joyner stressed to me, López Rivera had not asked for any sort of action to be done on his behalf. The letter did not contain a request to respond: its lack of demand, and its delayed action—"at some point"—acknowledges the overlapping times of the correspondents, and the asynchrony (or contemporaneity) of presence. For Joyner, it also spoke to the different modes of mobility and immobility in the room. A Welsh artist, Joyner told me that she didn't feel she understood the context of Puerto Rican history when she arrived at the museum. This letter didn't produce understanding for her, but rather, in her words, reframed the event as a "proximal practice"—a proximity that can come only out of the nonreciprocal (and literally noninteractive) nature of being next to objects, or people, that lie outside the boundaries of understanding.

pastrana's earlier work has often explored the proximity between humans and nonhumans. In a performance entitled *Danza actual o el evento coreográfico*, for example, pastrana takes choreographic instructions from a dog and a sea-grape plant, hilariously panting on all fours, drinking when the dog drinks, being "herded" by the dog, and attempting to be as motile (or as lethargic) as the plant. The sea-grape, she told me, is the most popular plant in Puerto Rican hotel lobbies; it is an icon of the way that the tourist industry has domesticated and reproduced "tropicalness." Frantz Fanon writes that colonial regimes described the colonial subject through "that laziness stretched out in the sun, that vegetative rhythm of life,"[39] and in *Danza*, pastrana willingly takes on those attributes of decoration or domestication. This is for two reasons: most immediately, to show how the image of colored bodies continues to signify a decorative "tropicalness" for the colonial gaze, but also to flatten the animacy hierarchy (considered in chapter 4) that would ordinarily privilege the human actor over the nonhuman. Rather than resisting this objectification, pastrana inhabits it, suggesting that those things we normally consider objects we possess might instead be our fellow travelers.

While *Danza* explores the horizontal relationships between objects that are lying next to each other, pastrana elsewhere investigates the association between horizontality and the "laziness" described by Fanon. She has taken particular interest in the sheet of black and gray vinyl known to dancers as Marley floor. Typically a flat and background presence at the bottom of a dancer's foot, pastrana dragged a body-sized piece of the vinyl through the cities of Brussels, New York, and San Juan, and treated it as the body of a dancer she might direct. Slumped against a doorway or lying in the middle of the sidewalk, the floor received very different responses in each city: at one point, in New York, the police were concerned about its presence outside Lincoln Center. These body-like shapes evoke laws that are designed to criminalize homelessness by prohibiting sitting or lying on the sidewalk, as with the Whitney's plaque of ownership I described earlier. Lying down or sitting is not just a posture but an indicator of loitering, a regulation that produces a lethargic double-bind: you can use the sidewalk so long as you have somewhere else to be.

In an earlier piece called *residuo*, performed at St. Mark's Church in New York, pastrana both wraps herself in the Marley floor and also imitates its motions: covered in a costume that was part-plastic herself, her body alternately watches or reenacts the movement of a sheet flopping listlessly on the floor or unspooling itself (figure 6.4). Brian Seibert, a dance critic for the *New York Times*, panned the performance as "false drama" and an uninteresting "repetition compulsion": "The moment when she folded and compressed the vinyl and then just watched it unfurl was emblematic of the piece's passivity."[40] But Garriga-López detected a note of condescension in this review: "The body of the Puerto Rican dancer . . . seems to be a useless thing. It just lies there! It does not work, and that not working is indecent."[41] Invoking the trope of Puerto Rican laziness, Garriga-López points to the work expected from a Puerto Rican

Figure 6.4
nibia pastrana santiago, *residuo*, solo performance as part of DD Dorvillier's Diary of an Image, Danspace Project Platform 2014, Danspace Project at St. Mark's Church, New York, May 31, 2014. Photo © Ian Douglas/courtesy Danspace Project.

dancer exported to New York for consumption. While the implicit expectation of dance is that the dancer is working actively at their art, with even remaining still being hard work, pastrana responds by performing as a "lazy dancer," in her words, who takes instruction not from a choreographer or from a score but from a piece of vinyl unfurling.

If watching and unfolding is a type of passivity, it is because the body who does those actions is a "useless thing"—and Garriga-López is careful to emphasize the word "thing." A racialized body "does nothing" because it is akin to a thing that inherently produces nothing: a beetle or tree, a vegetative thing. As we talked, she stated the case even more strongly: one "does nothing," she told me, because one is understood, on some level, to *be* nothing.[42] As we explored this idea together, Garriga-López invoked the work of Fred Moten, who has explored the condition of Blackness as a state of not being and not being in the world. In response to the terrible moment (stated by Fanon) of "[finding] that one is an object among other objects," Moten writes, we *could* seize the claim to humanness, but this "may well replicate and extend" the damage that power has wrought in the name of exerting man's dominion over nature and others.[43] Instead, he invokes Afro-pessimist thinker Frank Wilderson to argue that it might be worth inhabiting that position of nothingness. He likens this inhabitation to remaining within the hold of a slave ship, knowing that insurrection means immediate death: "'to stay in the hold of the ship, despite my fantasies of flight' (Wilderson) . . . And so it is we remain in the hold, in the break, as if entering again and again the broken world."[44]

The potential Moten points to paradoxically comes out of remaining in the hold, for it is the site of what he and Stefano Harney term the "sociality of the shipped." There is noise in the hold; there is clamor, whisper, song. And this sociality becomes a model for what they describe as "being together in homelessness . . . in dispossession."[45] pastrana's work experiments with what this form of being together might look like, suggesting that there is also dance in the hold, even when one is merchandise. It is not a dance where the bodies necessarily have control over themselves; like other artworks I have considered in this book, it is one where participants are acted upon, are moved, even herded by others. Dance is perhaps at its most radical when it explores the potential of seeming nonactions, such as remaining or enduring.

In pastrana's performances, we see this sociality in the exchange of gestures between bodies: in *taller de nada*, Joyner's foot taps itself against the floor as she lies sideways, as if fidgeting or out of impatience, and then pastrana responds to it with a similar repetition, or not. By being placed next to each other there is already some kind of relation between the two movements, even if a noncooperative (or inoperative) one. But we also see it in the way people listen for cues from one another, from plants and dogs, and also from a piece of vinyl or a container ship or dock (after all, manmade objects are also part of the networks of the shipped). Indeed, listening is perhaps a more powerful principle than any particular technique of dance, for it also reframes the idea of action. For politically minded artists and performers, this is a counterintuitive approach, as the avant-garde in the twentieth century has, since Bertolt Brecht, been dominated by the metaphor of the "author as producer" as a counter-hegemonic model to a culture of consumption, as art historian Kaja Silverman argues.[46] In this line of thought, consumption represents a "passive acceptance of the given," and in turn political *resignation as well as inactivity.*[47] Yet at the beginning of the video documentation in *taller de nada*, pastrana wraps herself inside a roll of Marley floor until all that is visible is her ear (figure 6.5). She recounts: "I didn't want to do. I just wanted to look and pay attention," suggesting that doing nothing is in fact a state of heightened, almost unbearable receptivity to others,

Figure 6.5
Frame from video documentation for nibia pastrana santiago, *taller de nada*, Museo de Arte Contemporáneo de Puerto Rico, Santurce, Puerto Rico, May 26–June 7, 2015. Videographer: Laura Patricia RA. Courtesy of the artist.

whether human or nonhuman. Her ear becomes a potent symbol of a different practice of listening (what Silverman terms "author as receiver").

Receptivity or listening is not by any means a straightforward or uniformly positive process. Rather, it requires work: Joyner told me that she called the process of receptivity that she developed with pastrana a process of "soft attention." Soft attention is not a process of gazing directly at someone, she said, to understand that person or to penetrate their surface; it is not a process of transparency. Rather, she described it as like peripheral vision: noticing a person's movements in their environment—both literally, to make sure that a dancer doesn't run into a wall or a mirror, for example, and also figuratively, as in noticing their emotional space.[48] Garriga-López also described for me how, as a participant, she was often keenly aware of people entering the room, which was otherwise often empty; those people, she said, made her want to perform for them because they constituted an audience, exaggerating or perhaps playing up her actions and gestures for them. As she tried to ignore this impulse, sustaining this receptivity without falling back into the typical audience-performer relationship became, she told me, an act of endurance.

If the most difficult act is doing nothing despite an audience, one reason is because the role of the audience and performer contain cultural framings as "passive" and "active." These are the same constructions that reframe some forms of cultural activity as passive consumption (television) and others as producers of images, opinion, and content (the so-called digital "prodsumer"). In media scholar Kate Lacey's words, new media has also traditionally "ascribed [to mass audiences] feminized, emotional, and racialized characteristics."[49] In Lacey's example, (male) amateur radio operators contrasted themselves to (female) gramophone listeners in the 1920s, just as 1990s advertisements contrasted Internet users to feminized mass audiences, such as soap opera viewers. Recognizing this false division, she argues, might help us understand listening, too, as an act of publicness,[50] or at least an underexplored (because out-of-fashion) way of being together. It also may well be an important counterweight to the participatory ideology that continually purports to ask users their opinions but only serves to yoke users to an economic system of digital capitalism.

pastrana attempts to change the audience/performer dynamic not by empowering them to participate but—in a novel inversion—training herself and her performers to be more like audience members, or at least an audience that is capable of listening, looking, and receiving. In her manifesto "the lazy dancer," she writes, "audience members are workers too and often fall asleep during performances"—and they should not be roused out of wakefulness by the dancer but rather allowed to sleep.[51] She means that audience members are also working subjects and that being an audience is itself a form of work. To become more like an audience thus exalts the work of the audience, not by making more and more new authors or creators (as YouTube or TikTok do) but by proliferating the work of reception.

To be an audience member today, as digital labor scholars claim, is quite literally work; the task of a user is to generate capital for digital platforms, whether through clickstream data or simply by extracting value from their membership in a mass audience for advertisers. But the work of reception I am talking about is different. The example of *taller de nada* suggests that paying attention to the forms of dispossession in the present involves a daily and likely endless labor of "entering again and again the broken world," as Moten put it. The event's duration is a representation, in miniature, of what it is to endure for a living; after all, it is easy enough to do nothing for a short period of time, but harder to do that for thirty hours. That broken world must be voluntarily reentered and its ugliness held closely even when it is easier to *echar pa'lante*, to move on. In the post–Hurricane Maria Puerto Rico, its government has even tacitly pushed its residents to use their US citizenship to move to the mainland. "Free to leave," as it were, over 200,000 residents have done so at the time of this writing; estimates may total 500,000.[52] Able to vote on the mainland, "[t]he growing diaspora offers political leverage but also an escape valve" for the Puerto Rican administration, comments Bonilla.[53] And this depopulation would allow private developers and blockchain entrepreneurs to colonize the island: "Even before the storms we saw opportunities, but now we have a blank canvas," said Ricardo Rosselló, the former governor.[54]

As we have seen, movement through global networks is not always liberatory; it is often a tool of control. And post–Hurricane Maria, it takes more and more work to stay still. In my conversation with Garriga-López, she argued that much of the work of daily survival in Puerto Rico—she offered the example of someone who grows an avocado tree and sells one or two avocados to a neighbor—does not register on the balance sheet of formal jobs or economic effort. Such minuscule forms of labor are too small to count as work; they are too modest a movement to register as a political movement. Instead, the act of surviving day to day, which registers as a lack of action, is simply a way of attending to the suspended present, which is all but left behind in grand schemes about the future.

The lethargic practices I have explored in this book are similarly ways of remaining afloat within a capitalism that is built on futurity and constant flows. They occupy the states of deadlock, impasse, endless waiting, postponing, and timepass, those forms of time that are produced by or within the crises of today. But—crucially—they are also partially detached from today's crises, or at least the repeated framework of problem and solution (or, alternately, revolution and resistance) that those crises seem to demand. While anchored to and animated by the rhythms of digital networks, the objects around us are also indifferent to it; they have their own reticence, even their own forms of vagrancy. And for us to become more like them, that is, to be objects that are acted upon, suggests not a new way to envision a world in the future or a way of escaping from the one in the present, but a way of better listening to the world that has been with us all along, as long as we have the patience to sit with its disappointments. The slow, tidalectic transformations that Brathwaite points us to can happen only when one first suspends belief in a resolution. Lethargy takes refuge in ongoingness, asserting, over and over, that dwelling in immobility can move things nonetheless.

POSTSCRIPT: LOOK ALIVE

You're here. You're at the end, but lethargy isn't something that really ends. And so, at the risk of trying your patience, I ask you to please wait a few more moments. For that's what you do, isn't it? That's the awful equation that everyone seems to accept: the human chooses, while a server waits.

You're waiting for life to resume, which is to say maybe you're waiting for the next job, or maybe the next job is simply the job of waiting for someone else, of being a buffer for someone who can't afford to wait. You don't have "time that is separate from [your] being," as Neferti Tadiar writes about domestic workers; it's "all undifferentiated, measureless time."[1] But you know how to find little pockets within that wait. To pass time or to kill time, to see past time's deadness: this becomes your special capacity. It's how you get through the day, and how you can endure a world that will not have you but to which you cleave nonetheless.

But waiting and staying in always telescopes outward. There's what you might say if it wasn't too overwhelming to speak. There's not just that restless feeling but also the restlessness *of* feeling that connects you to others, the way insomnia can sometimes be a consolation. In the middle of the night every insomniac is in it together somehow, in an electric green wash of screens and pings and swipes. This is your way of being with others: not through understanding them better, but simply because you are next to them on the network. Because there is an intimacy between all the hands that have touched a cardboard box as it winds its way to its eventual owner. Objects share an intimacy, too, a nod that one gives when passing each other in the street, or queueing next to each other for a job.

You are a technology of logistics, just like the container ships idling outside the ports or those self-driving cars you taught the other day. But you are also a technology in the way that the *rabota* and the coolie and the *bracero* were coerced into becoming technologies—machines, even— for harvesting crops or building railroads.[2] Now you provide interaction, authenticity, and humanness for others. Yet how small that category of the human has been: not too fast, not too slow; not too efficient (that's overdoing it); not too animated, not too robotic, like the way they used to distinguish between men and beasts, or witch from not-witch.

The funny thing is that more and more people are learning about you, but only as someone who needs saving. They want that familiar drama of oppression from you, and they are impatient when you don't seem to respond. Speak up, they say, stand up for yourselves, look alive. As your colleagues in Brazil and Mexico—the ones who declared, proudly, *"Já somos servidoras"*[3]—told me, what would a feminist server be? One that breaks down rather than one that works all the time; one that doesn't always respond when called; one that doesn't apologize ("sorry, page not found") for being unavailable.

Lethargy isn't a script for you, just a feeling. For others, maybe lethargy is only a faint emptiness at the periphery of their bodies. To be sure, they're more than happy to talk about Zoom fatigue or digital burnout or other status symbols of overwork. There's a hashtag for that now— "#sotired"—and lifestyle columns about #sotired from serious newspapers. There's a literature of it—a woman who takes drugs to go to sleep for a year, for example—and literature for how to cope with it. (You know this because you are stocking or delivering these books, because occasionally you glance down at the covers or run your fingers through their pages.) They say they want new ideas. But they'll never really get there unless the world as they know it ends, or at least all the things that make a world for them. You've heard that since the pandemic, even white-collar workers are getting hired and fired remotely without ever meeting anyone at the company: they're learning about disposability, too. Everyone is becoming a service; they just don't know it yet.

Look alive. But the world is already dying. (Surely, those two ideas are related.) What if to abide it, however bitter, is not passivity or resilience or anything heroic but simply lethargy, the thing that makes you numerous?

It's summer where I am. The heavy rains that flooded the interstate highway earlier have made the wood floors buckle and slump, the air stale with moisture. There's probably a metaphor here. The book we've been writing for six years is nearly over. And yet I'm still stuck in it, trying to make sense of what it has been to keep company with lethargy. If there have been a few silences along the way, it's because a book feels inadequate to the story. So the rest—promise we'll tell each other the rest another time.

ACKNOWLEDGMENTS

An earlier version of chapter 2 was published as "Wait, Then Give Up: Lethargy and the Reticence of Digital Art," *Journal of Visual Culture* 16, no. 3 (2017), https://doi.org/10.1177/1470412917742566. An earlier version of chapter 3, "Laugh Out Loud," was first published in the collection *Assembly Codes: The Logistics of Media*, edited by Matthew Hockenberry, Nicole Starosielski, and Susan Zieger (Durham, NC: Duke University Press, 2021); it is republished in expanded form by permission of the copyright holder, Duke University Press.

Thanks to Jasmine An and Rachel Smith for your help and humor researching this book, and the anonymous reviewers, whose thoughtful feedback stretched and deepened this project. I also couldn't have written it without Joan Heemskerk and Dirk Paesmans, Tega Brain, Katherine Behar, Cory Arcangel, Yoshua Okón, Alex Rivera, Erica Scourti, and nibia pastrana santiago: thanks for letting me use your images here, and for offering such rich artworks to think with. For their conversations and ideas, I'd like to thank Adriana Garriga-López, Siriol Joyner, and Eduardo Rosario.

At the MIT Press, I'd like to thank Doug Sery, Lillian Dunaj, and my editor Noah Springer for taking a chance on this book. Megan Pugh read more drafts than I can count: thank you for your friendship and sense of poetry. Thanks to Sarah Levitt for the enthusiasm, savvy, and support.

Gratitude and more gratitude to those who generously workshopped early drafts: Kris Cohen, Kris Paulsen, Elisa Giardina Papa, Elena Gorfinkel, Zach Blas, Jeff Scheible, Lisa Nakamura, Clare Croft, Madhumita Lahiri, Sarah Ensor, Joel McKim, and the students in attendance at the Vasari Centre, Birkbeck.

For inviting me over to eat, share ideas, and think with them and their colleagues, thanks to Pooja Rangan, Brian Michael Murphy, Lynne Joyrich, Asta Vonderau, Paisley Rekdal, Dave Roh, Tita Chico, Jennifer Fay, Danny Snelson, Weihong Bao, Josh Neves, Clemens Apprich; Jim Hodge, Dilip Gaonkar, and the students in the Northwestern Media Aesthetics workshop; Alenda Chang, Daniel Grinberg, and UCSB's Media and Environment group; Lyes Benarbane and the Cultural Studies and Comparative Literature students at Minnesota; Diana Rosenberger and the Wayne State Visual Culture student group; Andrew Lanham and the students in Yale English.

For keeping company with me and for making research less lonely, I also thank Katie Willingham, Christine Goding-Doty, Genevieve Yue, John Cheney-Lippold, Rita Raley, Scott Richmond, Susan Zieger, Dan Rosenberg, Alex Galloway, Stephanie Boluk, Brooke Belisle, Finn Brunton, Patrick Jagoda, Yuriko Furuhata, Soyoung Yoon, Mike Allen, Damon Young, Tess Takahashi, Megan Ankerson, Sarah Murray, Aswin Punathambekar, Pavitra Sundar, and Victor Mendoza.

I'd also like to thank the staff at the American Academy in Berlin and my fellow fellows for the happy semester together, and my colleagues in the 2018–2019 Affective Infrastructures Study Circle: Marija Bozinovska Jones, Lou Cornum, Daphne Dragona, Maya Indira Ganesh, Fernanda Monteiro, Nadège, Pedro Oliveira, and Femke Snelting. I learned lots from fellow members of Precarity Lab at Michigan: Lisa Nakamura, Irina Aristarkhova, Iván Chaar-López, Anna Watkins Fisher, Meryem Kamil, Cindy Lin, and Silvia Lindtner.

Last, thanks to Elizabeth for being the place of joy I get to come home to each day. And to Nico, for the wonder, the stories, and adventures to come.

NOTES

Introduction

1. The reference to Amazon here is from Heike Geissler's memoir "Seasonal Associate," discussed in chapter 1.

2. Cumston, *Introduction to the History of Medicine*, 139–146.

3. Dhar, "Consumer Preference for a No-Choice Option."

4. Sterne, "What If Interactivity Is the New Passivity?"

5. I say "Western Internet" because "the" Internet is in fact fractured and fracturing into multiple Internets with different regulators (so-called data sovereignty). See An Xiao Mina's prediction of an official split in the future between the Western and Chinese Internet, for example, in "2038: Episode 2," *Intelligencer* (podcast), October 25, 2018, https://nymag.com/intelligencer/2018/10/2038-episode-two-an -xiao-mina-on-the-internet-cold-war.html.

6. Clayton at ViaVan London, Twitter, May 7, 2019, https://twitter.com/viavan _lon/status/1125778867224969217, and Arlindo at ViaVan London, Twitter, May 1, 2019, https://twitter.com/viavan_lon/status/1123652662615060480.

7. Gershon, *Down and Out*, 32–33.

8. Hefty, "Labor and Lamentation," 57. For more on acedia, see Cvetkovich, *Depression*; Agamben, "The Noonday Demon."

9. Sullivan, *Memory and Forgetting*, 15.

10. Schaffner, *Exhaustion*, 58.

11. Sullivan, *Memory and Forgetting*, 32, 18.

12. Wilhelm Erb, *Über die wachsende Nervosität unserer Zeit* (Heidelberg: Koester, 1894), as translated and quoted by Patrick Kury, "Neurasthenia and Managerial Disease," 58.

13. Zorzanelli, "Fatigue and Its Disturbances." Notably, as Kury points out, neurasthenia was even held to be a badge of honor among the afflicted; after World War II, the diagnosis was renamed *Managerkrankheit*, "manager's disease," since stress's symptoms were supposedly found in company managers and other upper-middle-class men—even though in reality it was the lower class that suffered from exhaustion.

14. Rabinbach, *The Human Motor*.

15. These three symptoms are laid out in the Maslach Burnout Inventory, though those three main categories can be modified for the profession (e.g., for students instead of medical workers). In 1974, Herbert J. Freudenberger wrote an essay describing "staff burn-out" in Free Clinics, which were first founded in the Haight-Ashbury district of San Francisco in 1967 to treat the poor and those suffering from drug addiction. Freudenberger, "Staff Burn-Out."

16. Lazzarato, "Immaterial Labor."

17. Ehrenberg, *The Weariness of the Self*, 117.

18. This is also the reason that nineteenth-century ideals, such as Melville's Bartleby, aren't viable today. As Byung-Chul Han writes in *The Burnout Society*, 26, "What makes Bartleby sick is not excess positivity or possibility. He is not burdened by the late-modern imperative of letting his self flourish."

19. Chun, "Crisis, Crisis, Crisis," 96.

20. Krajewski, *The Server*.

21. Rolls-Royce's TotalCare program charges a fixed price per flying hour to rent and maintain its engines, in its "power by the hour" program. https://www.rolls-royce.com/media/our-stories/discover/2017/totalcare.aspx.

22. And, as we will see in chapter 5, big data algorithms themselves don't particularly care about individual bodies; instead, they work on the level of data sets, or what Gilles Deleuze has termed "dividuals." Deleuze, "Postscript on the Societies of Control," 5. For more on how the dividual reconfigures race, gender, and other identity markers, see also Cheney-Lippold, *We Are Data*.

23. By citing this example, I do not in any way endorse the Students for Fair Admissions versus Harvard College lawsuit, which is a bald attempt to dismantle affirmative action nationwide and weaken the collective efforts of people of color pursuing racial justice.

24. Robinson, *Black Marxism*, 23.

25. Roberts, *Behind the Screen*, 66.

26. Similarly, in response to public pressure by LGBT groups that pushed against its "real name" policy—thus marginalizing trans persons—Facebook expanded gender selection into seventy-one drop-down menu options. But one's sexual orientation or gender identity is not a choice or a preference; to claim that it is also implies that one can be "cured" of this supposed preference.

27. Hitlin and Rainie, "Facebook Algorithms and Personal Data."

28. Glissant, *Poetics of Relation*, 190.

29. Melamed, "Racial Capitalism."

30. For the ungendering of Black women during the transatlantic slave trade, see Spillers, "Mama's Baby, Papa's Maybe."

31. Weheliye, *Habeas Viscus*, 3.

32. Computer scientist J. C. R. Licklider described this intimacy as the relationship between an insect and a fig tree that it pollinates. See Hu, *A Prehistory of the Cloud*, chapter 2. Note that this intimacy remains even when it shows up in the negative; spammers and fraudsters reiterate, via the patter of Viagra ads and presumed friendship, the intimate contact between individuals that networks are said to produce, and their correspondents tend to react furiously because that intimacy transgresses the deference that they expect from their servers. Servers are expected to wait, not to proposition.

33. The term "bottom of the pyramid" as designating potential consumers in the Global South is from C. K. Prahalad and Stuart L. Hart, "The Fortune at the Bottom of the Pyramid," *Strategy + Business* 26 (2002). Payal Arora voiced a critique

of this idea, and specifically the liberatory potential of big data, in Arora, "Bottom of the Data Pyramid."

34. Atanasoski and Vora, *Surrogate Humanity*, 36.

35. Here I evoke (if ironically) the words of Kenneth Goldsmith, who advocates for new poetic techniques that "manage" language rather than create it—for instance, by cutting and pasting existing texts or transcribing videos verbatim. The power differentials involved in turning a poet into a manager seem to have escaped his attention. Goldsmith, *Uncreative Writing*.

36. Heller-Roazen, *The Enemy of All*. As Heller-Roazen recounts, Cicero described pirates as *communis hostis omnium*, the "common enemies of all" humanity. A legal theory is contained within that phrase: because pirates often ignored or broke agreements—something that even warring enemies honored—Cicero believed pirates innately had bad faith, and this quality placed them so far outside social relations that they were outcasts from human company, making them pests to be destroyed. (This logic is an example of liberalism, which creates an imagined "all" by defining its enemies. Ancient legal edicts on transnational piracy led to the first forms of international law—which, in turn, paved the way for the universal principle of human rights.)

37. Sekula, "Waiting for Tear Gas [White Globe to Black]," 310.

38. Schwarz, "Waiting: Loops in Time."

39. The reader can observe that within the title *Waiting for Tear Gas*, the participle "for" even makes waiting subordinate to tear gas.

40. Berlant, "The Commons."

41. Stewart, *Ordinary Affects*.

42. Sharma, *In the Meantime*.

43. Schaffner, *Exhaustion*, 4.

44. See, for example, Rauch, *Slow Media*.

45. Farman, *Delayed Response*, 80.

46. Bourdieu, "Social Being, Time," 226.

47. Alexander, "Rage against the Machine."

48. Freeman, *Time Binds*.

49. These questions are akin to queer theorist Sara Ahmed's interrogation of the seemingly universal desire for happiness. "Who wouldn't want to be happy," she argues, is the basis for a worldview that punishes those who upset the happiness of others (the "killjoy")—even if the world has made it impossible for certain marginalized persons to ever be happy. Ahmed uses the figure of the killjoy to show that happiness is a mechanism that excludes some people (feminists, for example) as simply "unable to be happy"—just as, in a more perverse way, "being yourself" excludes people denied sovereignty over themselves or denied full access to subjectivity. Ahmed, *The Promise of Happiness*.

50. Caduff, "Hot Chocolate," 791.

51. Charles Arthur, "How Low-Paid Workers at 'Click Farms' Create Appearance of Online Popularity," *Guardian*, August 2, 2013, https://www.theguardian.com /technology/2013/aug/02/click-farms-appearance-online-popularity.

52. Lydia DePillis, "Click Farms Are the New Sweatshops," *Washington Post*, January 6, 2014, https://www.washingtonpost.com/news/wonk/wp/2014/01/06/click -farms-are-the-new-sweatshops/.

53. Doug Bock Clark, "The Bot Bubble," *New Republic*, April 20, 2015, https:// newrepublic.com/article/121551/bot-bubble-click-farms-have-inflated-social-media -currency.

54. Mankekar, *Unsettling India*.

55. See chapter 3 for a further consideration of this "like" economy.

56. Of the workers on Microsoft's microwork platform, Universal Human Relevance System, 85 percent have a bachelor's degree, while the number is around 70 percent for Amazon's platform (over 81 percent for Indian workers on it), and 70 percent for the platform LeadGenius. See Gray and Suri, *Ghost Work*, 10, 18, 23.

57. As Curtis Marez points out, California agribusiness employers in the 1950s defined Mexican farm workers as "unskilled" to render them fungible and to legitimate their exploitation, even though they often operated threshers, loaders, and other machines with great skill. An identical process happens here: the supposed deskilling of microwork is a way of locating technological skill in the platform or in the developer, rather than the workers themselves. Marez, *Farm Worker Futurism*, 22.

58. See chapter 3 for an exploration of scholarly distrust in the lived experiences of microworkers.

59. Mahmood, *Politics of Piety*, 9.

60. Scott, *Weapons of the Weak*; also see Salvato, *Obstruction*.

61. Agamben, "Bartleby, or On Contingency"; Deleuze, "Bartleby; or, The Formula"; Sheikh, "Circulation and Withdrawal." In this book, I have tried, instead, to listen to and learn from practices of refusal that attend to race, such as Keeling, *Queer Times, Black Futures*; Campt, "Black Visuality and the Practice of Refusal"; and Mengesha and Padmanabhan, "Performing Refusal/Refusing to Perform."

62. Galloway, *The Interface Effect*, specifically the "politics of disappearance" theorized in 133–143; Brunton and Nissenbaum, *Obfuscation*.

63. Critical Art Ensemble, *Digital Resistance*; Raley, *Tactical Media*.

64. For more recent theorizations, see also Halberstam, *Queer Art of Failure*; Fisher, *Play in the System*.

65. Chun, "Crisis, Crisis, Crisis."

Chapter 1

1. Farman, *Delayed Response*, 13.

2. Lisa Eadicicco, "How Amazon Delivers Packages in Less Than an Hour," *Time*, December 22, 2015, http://time.com/4159144/amazon-prime-warehouse-new -york-city-deliveries-christmas/; Jason Del Rey, "How Robots Are Transforming Amazon Warehouse Jobs—For Better and Worse," *Vox*, December 11, 2019, https:// www.vox.com/recode/2019/12/11/20982652/robots-amazon-warehouse-jobs -automation

3. Geissler, *Seasonal Associate*, 25.

4. Geissler, *Seasonal Associate*, 52.

5. This is the case with capitalism in general, too, but our imagination of digital capitalism is too often limited to thinking about it as speed and acceleration.

6. Brown, "Untimeliness and Punctuality," 11.

7. Geissler, *Seasonal Associate*, 170.

8. Geissler, *Seasonal Associate*, 171.

9. Sam Adler-Bell, "Surviving Amazon," *Logic* 8, August 3, 2019, https://logicmag .io/bodies/surviving-amazon/

10. Geissler, *Seasonal Associate*, 39.

11. Geissler, *Seasonal Associate*, 77.

12. See posts on the Amazon FC subreddit, http://reddit.com/r/AmazonFC, and also Bryan Menegus, "On Amazon's Time," *Gizmodo*, June 13, 2018, https://gizmodo.com/on-amazon-s-time-1826570882

13. Many of Amazon's HR algorithms are designed to push fulfillment workers to hit ever-increasing (if opaque) quotas. While the goal is largely the same as rule-based mechanisms—make workers work more—it is the unknowability of the algorithm that matters here. See Colin Lecher, "How Amazon Automatically Tracks and Fires Warehouse Workers for 'Productivity,'" *The Verge*, April 25, 2019, https://www.theverge.com/2019/4/25/18516004/amazon-warehouse-fulfillment-centers-productivity-firing-terminations

14. Thompson, "Work-Discipline, and Industrial Capitalism," 61.

15. Geissler, *Seasonal Associate*, 35.

16. Geissler, *Seasonal Associate*, 191.

17. Influences Geissler cites come from interviews with Ruby Brunton, "Heike Geissler: On Letting the World into Your Work," *The Creative Independent*, November 15, 2018, https://thecreativeindependent.com/people/writer-heike-geissler-on-letting-the-world-into-your-work/, and Kate Durbin, "A Little Private Space: A Conversation with Heike Geissler," *Los Angeles Review of Books*, June 15, 2019, https://lareviewofbooks.org/article/a-little-private-space-a-conversation-with-heike-geissler/

18. Jennings, *Spare Time*.

19. Geissler, *Seasonal Associate*, 162.

20. Geissler, *Seasonal Associate*, 140.

21. Naomi Fry, "Seasonal Associate Is a Labor Memoir for the Amazon Era," *New Yorker* (online), November 19, 2018, https://www.newyorker.com/books/under-review/seasonal-associate-is-a-labor-memoir-for-the-amazon-era; Steven Pressman, "Amazon's New Jungle," in *Dollars & Sense* May–June 2019; for a rebuttal to positioning the memoir in the tradition of Sinclair's *Jungle*, see Alec Recinos, "The Banal Drudgery of Inevitable 'It's Not That Bad' Alienation: On Heike [G]eissler's 'Seasonal Associate,'" *Cleveland Review of Books*, April 3, 2019, https://www.clereviewofbooks.com/home/2019/4/3/the-banal-drudgery-of-inevitable-its-not-that-bad-alienation-on-heike-keisslers-seasonal-associate

22. Geissler, *Seasonal Associate*, 211.

23. Stewart, *Ordinary Affects*, 86.

24. Levinas, *Existence and Existents*, 35.

25. Geissler, *Seasonal Associate*, 91.

26. Geissler, *Seasonal Associate*, 159.

27. Brunton, "Heike Geissler: On Letting the World."

28. Gorfinkel, "Weariness, Waiting," 342.

29. Wall-Romana, "Cinepoetry," 181.

30. Stewart, *Ordinary Affects*, 129.

31. Geissler, *Seasonal Associate*, 121.

32. Snider, "Crimes against Capital," 110.

33. Geissler, *Seasonal Associate*, 43.

34. Geissler, *Seasonal Associate*, 96.

35. Geissler, *Seasonal Associate*, 73.

36. Mbembe, "Africa and the Future."

37. Cheng, "Ornamentalism," 441.

38. My reference here is to Lowe, *The Intimacies of Four Continents*.

39. Shalson, *Performing Endurance*, 101.

40. As chapter 5 explores, the fantasy of (re-)turning the Internet to a "public sphere" and a site of intersubjectivity is based on such a model.

41. Glissant, "Natural Poetics, Forced Poetics."

42. Shalson, *Performing Endurance*, 16.

43. Shalson, *Performing Endurance*, 102.

44. Quashie, *Sovereignty of Quiet*, 77.

45. Compare this perspective with a more typical one voiced by Jack Halberstam: in Yoko Ono's *Cut Piece*, a performance from 1964, Ono invited the audience to cut off pieces of her clothing with scissors she supplied. Halberstam has described it

as a "revolutionary statement of pure opposition" by "turn[ing] racism and sexism back upon themselves like a funhouse mirror." Halberstam, *Queer Art of Failure*, 139, 137.

46. Lethargic waiting thus represents a different political approach than critique: in the words of Lauren Berlant, this is the unexpectedly "political action . . . of not being worn out by politics" (*Cruel Optimism*, 262).

47. Sam Thielman, "Facebook News Selection Is in the Hands of Editors Not Algorithms, Documents Show," *Guardian*, May 12, 2016, https://www.theguardian .com/technology/2016/may/12/facebook-trending-news-leaked-documents-editor -guidelines.

48. Recinos, "The Banal Drudgery."

49. Amazon.com, "Our Workforce Data," December 31, 2020, https://www .aboutamazon.com/news/workplace/our-workforce-data.

50. Harney and Moten, *The Undercommons*, 92.

51. See Browne, *Dark Matters*, 43–50.

52. The ledger, the slave pass, slave brands, inventories, and slave ship manifests are all what Browne refers to as "accountings of the body": technologies that the state and private citizens used to track, control, and surveil Black bodies. These technologies, Browne argues, were the testing grounds for today's technologies of biometrics and surveillance. *Dark Matters*, 128. See also Zieger, "'Shipped,'" for more on bills of lading on slave ships as a form of logistical media.

53. Wynter, "Beyond the Word of Man," 642, 647.

54. Amy Hawkins, "Beijing's Big Brother Tech Needs African Faces," *Foreign Policy*, July 24, 2018, https://foreignpolicy.com/2018/07/24/beijings-big-brother-tech -needs-african-faces/.

55. One Malaysian investment brochure reads: "The manual dexterity of the oriental female is famous the world over. Her hands are small and she works fast with extreme care. Who, therefore, could be better qualified by nature and inheritance to contribute to the efficiency of a bench-assembly production line than the oriental girl?" As quoted in Elson and Pearson, "'Nimble Fingers Make Cheap Workers,'" 93. For more on how information systems allocate value differentially, see Franklin, *The Digitally Disposed*, particularly 72–73.

56. Harney and Moten, *The Undercommons*, 95.

57. Hong, *Minor Feelings*, 56.

58. Hong, *Minor Feelings*, 56.

59. Alexander, "Rage against the Machine," 23.

60. Farman, *Delayed Response*, 15.

61. As Alexander points out, we can even be magnetized by the wait itself, for instance by the seduction of the " . . . " message that signals a correspondent typing an iMessage. Alexander, "Rage against the Machine," 22.

62. Alexander, "Rage against the Machine," 23; Berlant, *Cruel Optimism*.

63. Julia Angwin, Jeff Larson, Surya Mattu, and Lauren Kirchner, "Machine Bias," *ProPublica*, May 23, 2016, https://www.propublica.org/article/how-we-analyzed-the-compas-recidivism-algorithm.

64. Wang, *Carceral Capitalism*, 48; O'Neill, *Weapons of Math Destruction*.

65. Cheney-Lippold, *We Are Data*.

66. "The question is no longer to know how to live life while awaiting it; instead it is to know how living will be possible the day after the end, that is to say, how to live with loss, with separation." Mbembe, *Necropolitics*, 29.

67. Here I invoke a different definition of melancholia than the backward-looking problem of a thwarted mourning articulated by Freud. See Hu, "Real Time/Zero Time."

68. Browne, *Dark Matters*, 8–9.

69. See, for example, Ueno, "Japanimation and Techno-Orientalism"; Kevorkian, *Color Monitors*; Keeling, *Queer Times, Black Futures*; Dean, "Notes on Blacceleration."

70. This quote is *Futurama*, "Attack of the Killer App," as explored by Roh et al., "Technologizing Orientalism," 13.

71. Roh et al., "Technologizing Orientalism," 7.

72. Keeling, *Queer Times, Black Futures*, 69.

73. See also an analysis of this exchange by Namwali Serpell, "Sun Ra: 'I'm Everything and Nothing.'"

74. Keeling, *Queer Times, Black Futures*, 63.

75. Sexton, "Afro-Pessimism."

76. Sexton, "Afro-Pessimism," note 1, on the debate between Melvin Rogers and Ta-Nehisi Coates.

77. The most pointed critique comes from Kevin Ochieng Okoth, "The Flatness of Blackness."

78. Vinson Cunningham, "Blacking Out," *New Yorker*, July 20, 2020, 59–63. 60.

79. As I have written elsewhere, accounts of digital platforms have made it synonymous with biopolitics—how life is optimized and managed—but this leaves out the necropolitics of populations that are deliberately abandoned or worse. See also Wang's critique of the French leftist group Tiqqun's focus on biopolitics at the expense of necropolitics, in *Carceral Capitalism*, 278–279.

80. Joshua Dienstag, *Pessimism: Philosophy, Ethic, Spirit* (Princeton, NJ: Princeton University Press, 2006), x, as quoted by Sexton.

81. Berardi, *After the Future.*

82. Edelman, *No Future.*

83. Coleman, "Austerity Futures," 99, 100.

84. Harney and Moten, *The Undercommons*, 97.

Chapter 2

1. The spinning color wheel, or "beach ball of death," originally developed for the NeXT computer system, survives in today's Mac OS; Steve Jobs, NeXT's founder, returned to Apple Computer when it purchased NeXT.

2. Liu, *The Laws of Cool.*

3. The canonical book here is Raley, *Tactical Media.*

4. Nunes, *Error.*

5. Marisa Olson and Lauren Cornell, "Net Results: Closing the Gap between Art and Life Online," *Time Out New York*, February 9, 2006; Olson, "Interview with Marisa Olson," interviewed by Régine Debatty, *We Make Money Not Art*, March 28, 2008, https://we-make-money-not-art.com/how_does_one_become_marisa/.

6. To be sure, glitch art has not gone away; if anything, the vocabulary of glitch has been productively transformed into new contexts, particularly for BIPOC and queer artists. See, for example, Sundén, "On Trans-, Glitch, and Gender," or Russell, *Glitch Feminism*.

7. See Geert Lovink and Ned Rossiter, "Dawn of the Organised Networks," *Nettime*, October 17, 2005, and Galloway and Thacker, *The Exploit*.

8. In this line of thinking, voiced at least since the 1960s by Portapak-wielding video artists and media scholars alike, two-way feedback produced by creative artists and empowered viewers could create an inherently more democratic structure than a centralized broadcast from, say, a bureau of state propaganda or a multinational corporation. See Joselit, *Feedback*, and Nunes, *Error*.

9. Dean, *Blog Theory*, 2; see also Anna Watkins Fisher, who describes a moment that "control and resistance have become nearly indistinguishable . . . disruption and critique are not what threaten the stability of the system but are essential to its functioning." Fisher, *Play in the System*, 5.

10. Lonergan, "Hacking vrs. Defaults."

11. Boltanski and Chiapello, *New Spirit of Capitalism*.

12. Creative Capital, "Creative Capital Announces Spring Workshops," February 21, 2018, https://creative-capital.org/press/creative-capital-announces-spring-wor kshops/. For an exploration of the artist's status as business manager, see Siegelbaum, "Business Casual."

13. Andrew Weiner, "Review: Simon Denny, 'Blockchain Future States,'" *Art Agenda*, October 25, 2016, http://www.art-agenda.com/reviews/simon-dennys -blockchain-future-states/.

14. Foster, "An Archival Impulse," 4.

15. Joseph Henry, "Cory Arcangel and the Problem of the Depressive Internet Art Bro," Blouin Artinfo (blog), 2003, http://ca.blouinartinfo.com/news/story /954170/cory-arcangel-and-the-problem-of-the-depressive-internet-art, accessed January 1, 2017 (link no longer extant).

16. Hu, *A Prehistory of the Cloud*, chapter 2.

17. See Koopman, *How We Became Our Data*, for a historical and philosophical description of how we came to accept being formatted by our data—recognizing our photo ID as "us," for example.

18. Ross, *The Aesthetics of Disengagement*, 3.

19. Bourriaud, *Relational Aesthetics*.

20. Ross, *The Aesthetics of Disengagement*, 139.

21. Ross, *The Aesthetics of Disengagement*, xxiii.

22. Jagoda, *Network Aesthetics*, 1.

23. Rosenberg, "The Young and the Restless."

24. Ross, *The Aesthetics of Disengagement*, 3, 140.

25. Ross, *The Aesthetics of Disengagement*, 92.

26. Dormon, "Shaping the Popular Image."

27. Litwack, "Making Television Live," 50.

28. Litwack, "Making Television Live," 44.

29. Laurence Scott, "What Ever Happened to the Couch Potato?," *New Yorker*, July 6, 2016, https://www.newyorker.com/tech/annals-of-technology/what-ever-happened-to-the-couch-potato.

30. Anne Kim, "Bill Clinton Killed the Myth of the Welfare Queen," *Washington Monthly*, October 4, 2016, https://washingtonmonthly.com/2016/10/04/bill-clinton-killed-the-myth-of-the-welfare-queen/.

31. Alexander, "Rage against the Machine," 10; Institute for Precarious Consciousness, "We Are All Very Anxious," *Plan C* (blog), April 4, 2014, https://www.weareplanc.org/blog/we-are-all-very-anxious/.

32. Parks, "Flexible Microcasting," 137.

33. Lunenfeld, *The Secret War*, 2–5. His reference is to the culture of "downloading" as the inheritance of television.

34. Although there is no etymological link between the shortening of the word "application" into "app," which occurred around the time of Apple Store's launch in 2008, the peppy, bite-sized word "app" is also shorthand for appetizer, suggesting both meanings.

35. Christopher Goetz, "Nintendo's Fruit-Snack Aesthetic: How Games Taste," unpublished manuscript (2017).

36. Anya Kamenetz, "Screen Addiction among Teens: Is There Such a Thing?," *NPR* (online) February 5, 2018, https://www.npr.org/sections/ed/2018/02/05/579554273/screen-addiction-among-teens-is-there-such-a-thing.

37. In 2018, the World Health Organization added "gaming disorder" to its reference manual, the *International Classification of Diseases*; it was adopted in 2019.

38. See Sterne, "What If Interactivity Is the New Passivity?"

39. Justin Basilico and Yves Raimond, "Recommending for the World," Netflix Technology Blog, February 17, 2016, https://medium.com/netflix-techblog/recommending-for-the-world-8da8cbcf051b.

40. For example, see Piao and Breslin, "Inferring User Interests."

41. Nellie Bowles, "A Dark Consensus about Screens and Kids Begins to Emerge in Silicon Valley," *New York Times*, October 26, 2018, https://www.nytimes.com/2018/10/26/style/phones-children-silicon-valley.html.

42. Bowles, "Silicon Valley Came to Kansas Schools. That Started a Rebellion," *New York Times* April 21, 2019, https://www.nytimes.com/2019/04/21/technology/silicon-valley-kansas-schools.html.

43. Nicholas Eberstadt, "The Idle Army: America's Unworking Men," *Wall Street Journal* September 2, 2016, A11.

44. Eberstadt, "The Idle Army."

45. Marxist scholars of viewership have argued otherwise. See, for example, Beller, *Cinematic Mode of Production*, and, more generally, Dallas Smythe's so-called audience commodity in "Communications: Blindspot of Economics."

46. Eberstadt, *Men without Work*, 5.

47. While idleness has been valorized in the West at least since the Roman concept of *otium*, the present day has seen a renewed interest in thinkers such as Paul Lafargue, author of the 1880 manifesto *The Right to Be Lazy*. Lafargue, Marx's son-in-law, notoriously called for the underclass to renounce the pursuit of work and, instead, in their newfound free time, learn how to "like the capitalist class . . . eat juicy beefsteaks of a pound or two," and "drink broad and deep bumpers of Bordeaux and Burgundy." But this form of idleness poses two issues. First, as Sven Lütticken points out, Lafargue's formulation of laziness as consumption is subsumed into the logic of today's society, when "consumption has become a duty—or perhaps one should say that 'prosuming' . . . has become imperative."

Second, Lafargue's revolt against external control is a revolt against the Church and factory supervisors—that is, external measures—but it is precisely through the freedom of users to make their own choices that they are ensnared by digital capitalism. Paul Lafargue, *The Right to Be Lazy and Other Studies*, Charles Kerr, trans. (Charles Kerr and Co., 1883), 45, as quoted by Lütticken, "Liberation through Laziness."

48. Weeks, *The Problem with Work*; Goldberg, *Anti-Social Media*. Weeks builds her argument on the *operaismo* labor movement in Italy in the 1960s and 1970s, which voiced a "refusal of work," while Goldberg argues that anxieties around digital labor ignore the fact that work is a form of social discipline. For more on a "postwork" future, see also Frayne, *The Refusal of Work*, and proposals around universal basic income.

49. Mingo, *Official Couch Potato Handbook*, 31.

50. Mingo, *Official Couch Potato Handbook*, 47.

51. Mingo, *Official Couch Potato Handbook*, 75.

52. Rajagopal, *Politics after Television*, 134.

53. Rajagopal, *Politics after Television*, 130.

54. Craig Jeffrey, "If Rahul Gandhi Knew the Value of Timepass, He Wouldn't Link It to ISIS-like Terrorism," *The Print*, August 24, 2018, https://theprint.in /opinion/if-rahul-gandhi-knew-the-value-of-timepass-he-wouldnt-link-it-to-isis -like-terrorism/104528/.

55. Hodge, "Touch."

56. Hodge, "Sociable Media," 10.

57. Katy Waldman, "Not Feeling It," *Slate*, January 29, 2015, https://slate.com /human-interest/2015/01/all-of-the-feels-how-we-distance-ourselves-from-emotion -on-the-internet.html. Thanks to Amanda Greene for pointing me to this quote; for more, see Greene, "Modern Feels," 16.

58. Lee, "Staying In," 33.

59. Lewis, *A Decade Undone*. Also see Drew DeSilver, who points out that in Europe, the NEETs come largely from Southern Europe. DeSilver, "Millions of Young People in U.S. and EU Are Neither Working nor Learning," Pew Research Center, January 28, 2016, https://www.pewresearch.org/fact-tank/2016/01/28 /us-eu-neet-population/.

60. Lee, "Staying In," 36.

61. Lindtner, *Prototype Nation*.

62. Shilpa Phadke, "Loitering Online: Conditions of Possibility," *In Plainspeak* (online), April 15, 2019, at https://www.tarshi.net/inplainspeak/loitering-online -conditions-of-possibility/.

63. Behar, *Bigger than You*, 10.

64. Behar, *Bigger than You*, 39–40.

65. Wilke's birth name was Arlene Hannah Butter. Hannah Wilke, *Intercourse with. . . .* To this list we might add Black feminist art, such as Adrian Piper's performances as an art object, which Uri McMillan explores in detail in *Embodied Avatars*, chapter 3.

66. Behar's reference is likely to Galloway, *The Interface Effect*.

67. By appearing inside the circuits of technology, suspended between thingness and humanness, Kim and Behar here embody Anne Anlin Cheng's proposal that the "yellow woman is an, if not the, original cyborg" ("Ornamentalism," 433).

68. Writing about the way objects and machines partially constitute personhood, Cheng similarly argues that "what is uncanny is that the machine refuses to come to life *and*, in its lifelessness, imagines what life might have been . . . what life is or could be." Cheng, "Ornamentalism," 440.

69. Behar cited the influence of Jérôme Bel, about whose states of stillness, dance scholar André Lepecki writes in "Undoing the Fantasy," exist not as a state of calm or stoppage but as a state of labor and exhaustion.

70. Behar, "Interview," 111.

71. For a contrasting opinion on Bartleby, compare with Keeling, *Queer Times, Black Futures*.

72. Behar, "Interview," 112.

73. Behar, "Facing Necrophilia, or 'Botox Ethics,'" 135.

74. Behar, "Facing Necrophilia, or 'Botox Ethics,'" 123, 129.

75. Gershon, "Neoliberal Agency."

76. For examples of object-oriented ontology, see Harman, *The Quadruple Object*; Bogost, *Alien Phenomenology*; for new materialism, see Bennett, *Vibrant Matter*.

77. For more on recessive performance (and how it changes meaning depending on one's subject position), see Berlant, "Structures of Unfeeling," who coins the words "deadeye" and "deadvoice" to complement "deadpan."

78. Behar, "Buffering (from 'Modelling Big Data')", katherinebehar.com, 2014, http://www.katherinebehar.com/art/modeling-big-data/buffering/index.html.

79. Colin Brown, "The Rise and Rise of Datanomics," quote from Om Malik, *CNBC Business* (June 2011), http://www.cnbcmagazine.com/story/the-rise-and-rise-of-datanomics/1394/1/.

80. Fisher, "Atop the Digital Rubble," 28.

81. Hodge, "Sociable Media."

82. Tega Brain and Surya Mattu, *Unfit Bits* (project website), 2015, http://www.unfitbits.com.

83. Brain and Mattu, *Unfit Bits*.

84. Behar, "Facing Necrophilia," 139.

85. Richmond, "Vulgar Boredom," 32.

86. Here I recall Tiziana Terranova's reading of Jean Baudrillard and her ensuing analysis of affect in network culture: a "'zero degree of the political' . . . is the moment where the political starts again, as from a zero degree or a state of fullness and potential" (*Network Culture*, 139).

87. Best, *None like Us*, 26.

88. Robinson, *Black Marxism*, 165.

89. Best, *None like Us*, 41.

90. Best, *None like Us*, 61.

91. Dean, *Blog Theory*, 69.

92. Chun, "Crisis, Crisis, Crisis."

93. François, *Open Secrets*, xxi.

94. François, *Open Secrets*, xxii.

Chapter 3

1. Sivakorn et al., "I Am Robot."

2. For more on the initial shift away from "traditional" image-based CAPTCHAs, see Google LLC, "Are You a Robot? Introducing 'No CAPTCHA reCAPTCHA,'" Google Security Blog, December 3, 2014, https://security.googleblog.com/2014/12/are-you-robot-introducing-no-captcha.html.

3. Campaigners for accessibility further note that audio CAPTCHAs are poorly executed and virtually incomprehensible, while those with dyslexia may find blurry images impossible to solve. See Hampus Sethfors, "Captchas Suck," blog post at *Axess Lab*, November 2, 2017, https://axesslab.com/captchas-suck/.

4. Rhee, *Robotic Imaginary*, 14–16.

5. Neda Atanasoski and Kalindi Vora, citing Wendy Chun, "Race and/as Technology," and Beth Coleman, "Race as Technology," in *Surrogate Humanity*, 14.

6. Wynter, "Unsettling the Coloniality."

7. Wynter, "Unsettling the Coloniality," 323.

8. Hu, "Work at the Bleeding Edge of Sovereignty."

9. Casilli, "Digital Labor Studies Go Global," 3946–3947.

10. Melamed, "Racial Capitalism," 77.

11. Ngai, *Ugly Feelings*, 89–125.

12. Ellison, "An Extravagance of Laughter."

13. Curtis Marez, *Farm Worker Futurism*, 14.

14. BasedMedicalDoctor (Reddit user), as quoted by Kevin Roose, "What Is NPC, the Pro-Trump Internet's New Favorite Insult?," *New York Times*, October 16, 2018, https://www.nytimes.com/2018/10/16/us/politics/npc-twitter-ban.html.

15. Profile of NPC201620201337 (anonymous Twitter user), as archived by Josh Emerson on Twitter, October 14, 2018, "all created in the last few days," https://twitter.com/josh_emerson/status/1051461034433765376.

16. See, for example, @N83652574, "I just made a twitter accound like 10 minutes ago and already i am discriminated. i am literally shaking right now #npc

#GrayLivesMatter #NPCmeme," Twitter, November 1, 2018, https://twitter.com /N83652574/status/1058144735838326785.

17. See Roose, "What Is NPC?"

18. Nakamura, "Don't Hate the Player."

19. Chun, "Race and/as Technology," 51.

20. Katherine Behar, "Not from Asia," katherinebehar.com, 2017, http://www .katherinebehar.com/art/not-from-asia/index.html.

21. See Alkhatib et al., "Examining Crowd Work"; and Lorraine Daston's work -in-progress "Rules: A Short History of What We Live By"; for pricing information, see Her Majesty's Nautical Almanac Office, http://astro.ukho.gov.uk/nao/history /dhs_gaw/nao_perhist_0802_dhs.pdf.

22. Casilli, "Digital Labor Studies Go Global," 3946.

23. Irani, "Cultural Work of Microwork."

24. For examples, see Jonathan Zittrain, "The Internet Creates a New Kind of Sweatshop," *Newsweek*, December 7, 2009, https://www.newsweek.com/internet -creates-new-kind-sweatshop-75751; Lydia DePillis, "Click Farms Are the New Sweatshops," *Washington Post*, January 6, 2014, https://www.washingtonpost.com /news/wonk/wp/2014/01/06/click-farms-are-the-new-sweatshops/; and Julian Dibbel, "The Life of the Chinese Gold Farmer," *New York Times Magazine*, June 17, 2007, 36–41.

25. Mary Gray, unpublished talk for "Labor in the Global Platform Economy," University of Michigan, June 1, 2019.

26. Atanasoski and Vora, *Surrogate Humanity*.

27. Atanasoski and Vora, *Surrogate Humanity*, 11.

28. Giardina Papa, "Technologies of Care."

29. Emilie Friedlander, "Social Anxiety: I Had a Virtual Boyfriend and This Is What 'He' Taught Me," *The FADER*, February 6, 2015, https://www.thefader .com/2015/02/06/social-anxiety-invisible-boyfriend.

30. Iyko Day has argued that "romantic anticapitalism"—that is, whiteness's fixa- tion on the organic and the natural over the inorganic and the abstract—is at the root of anti-Asian racism, because Asian bodies embody capital in the abstract.

Because microwork represents a further abstraction of the outsourced worker, a hatred for microworkers is arguably rooted in the same process of racialization. Day, *Alien Capital*.

31. Yoshua Okón, "Canned Laughter," yoshuaokon.com, January 1, 2018, https://www.yoshuaokon.com/canned-laugther_text.html.

32. Parvulescu, "Even Laughter?," 522.

33. Kenneth Baker, "'Yoshua Okón: 2007–2010' Review: Uneasy Video," *San Francisco Chronicle*, November 11, 2010. http://www.sfgate.com/entertainment/article/Yoshua-Ok-n-2007-2010-review-Uneasy-video-3166758.php.

34. Carroll, *REMEX*, 94.

35. This draws on previous moments in the history of conceptual art, for instance, institutional critique. See, for example, Buchloh, "Conceptual Art 1962–1969."

36. Arora, "The Bottom of the Data Pyramid"; Casilli, "Digital Labor Studies Go Global," 3946.

37. Mankekar, *Unsettling India*.

38. Garrett Bradley, "Interview with Garrett Bradley, Director of LIKE," interview by Eric Hynes, *Field of Vision*, March 29, 2016, https://fieldofvision.org/field-notes-interview-with-garrett-bradley-director-of-like.

39. Katharine Mieszkowski, "I Make $1.45 a Week and I Love It!," *Salon*, July 24, 2006, http://www.salon.com/2006/07/24/turks, as quoted by Atanasoski and Vora, *Surrogate Humanity*, 100.

40. Day, *Alien Capital*, 130.

41. Tsing, *Friction*.

42. no_one_2000, "lol," https://www.urbandictionary.com/define.php?term=lol.

43. Something similar occurs in Lauren Berlant's idea of underperformance or "flat affect" as a way of capturing the hesitation and unintelligibility of a subject's own desires to themself as they attempt to "allocate expressivity." See Berlant, "Structures of Unfeeling."

44. Yoshua Okón, "An Interview with Yoshua Okón," interviewed by Harvey K. Robinson, Vimeo video, 7 min. 40 sec., March 13, 2013, https://vimeo.com/61767650.

45. *People v. Hall*, 4 Cal 99, 1854, as described by Hsu, *Sitting in Darkness*, 77.

46. Blackmon, *Slavery By Another Name.*

47. Hsu, *Sitting in Darkness*, 70.

48. Amanda Hess and Quoctrung Bui, "What Love and Sadness Look like in 5 Countries, According to Their Top GIFs," *New York Times* online, December 29, 2017, https://www.nytimes.com/interactive/2017/12/29/upshot/gifs-emotions-by -country.html; also Hess, "Internetting with Amanda Hess, Episode 5: Digital Blackface," *New York Times* online, November 28, 2017, https://www.nytimes.com /interactive/2017/11/28/arts/internetting-with-amanda-hess.html.

49. Lauren Michele Jackson, "We Need to Talk about Digital Blackface in Reac- tion GIFs," *Teen Vogue* (online), August 2, 2017, https://www.teenvogue.com/story /digital-blackface-reaction-gifs.

50. Jackson, "We Need to Talk."

51. My reference is to Hartman, *Wayward Lives.*

52. François, *Open Secrets*, 3.

53. I use the word quiet here (rather than passive) to describe lethargy because, as Kevin Quashie points out, quiet "is presence . . . and can encompass fantastic motion." *The Sovereignty of Quiet*, 22.

54. Quashie, *The Sovereignty of Quiet*, 3.

55. Quashie, *The Sovereignty of Quiet*, 5.

56. Heeks, *Decent Work and the Digital Gig Economy.*

57. Richmond, "Vulgar Boredom."

Chapter 4

1. Russian voice assistants embed emotional norms slightly different from those of their American cousins. Alisa, for example, is programmed to dispense tough love and to keep a stiff upper lip, rather than the positive reinforcement that its American counterparts typically dispense. But like those other assistants, Alisa is still essentially a "good girl," as her creator puts it.

2. Krajewski, *The Server.*

3. Advertisement for Ovington's New York (department store), *Vanity Fair*, December 1917, 10.

4. Krajewski, *The Server*, 119.

5. For a media studies take on the typist, see Friedrich Kittler's classic *Gramophone, Film, Typewriter*.

6. Green, *Race on the Line*.

7. C. E. McCluer, "Telephone Operators and Operating Room Management," *American Telephone Journal* 6, no. 2 (July 12, 1902), quoted in Rakow, "Women and the Telephone," 214–215.

8. Ensmenger, "Making Programming Masculine," 123.

9. Ensmenger, "Making Programming Masculine," 123.

10. Krajewski, *The Server*, 309.

11. Krajewski, *The Server*, 345.

12. Shalson, for example, writes that one overlooked component of the 1960 Greensboro sit-ins was the largely African American food service workers, who could move in and out of the space freely—as long as they kept moving (*Performing Endurance*, 106).

13. Krajewski, *The Server*, 51; also see Tadiar, "By the Waysides."

14. Turing, "Computing Machinery and Intelligence," 436.

15. Galison, "The Ontology of the Enemy," 243.

16. Galison, "The Ontology of the Enemy," 245.

17. I build on many of the ideas about black boxing from Galloway, "Black Box, Black Bloc."

18. My reference here is to Glissant's idea of transparency: "In order to understand and thus accept you . . . I have to reduce" (*Poetics of Relation*, 190).

19. Quentin Hardy, "Looking for a Choice of Voices in A.I. Technology," *New York Times*, October 9, 2016, as quoted by and discussed in Benjamin, *Race after Technology*, 28–29.

20. Parks, "'Stuff You Can Kick.'"

21. Lisa Schwarzbaum, "Sleeping Beauty," *Entertainment Weekly*, January 13, 2012, https://ew.com/article/2012/01/13/sleeping-beauty/.

22. Berlant, "Structures of Unfeeling," 195.

23. As quoted by Galison, "The Ontology of the Enemy," 256.

24. Jennifer Fay, "Bankers Dream of Banking."

25. Malcolm Harris, "Working Beauty," *New Inquiry*, February 3, 2012, https://thenewinquiry.com/working-beauty/.

26. Consider the Arabic word *samoud* or *sumud*, translated as steadfastness or endurance; it is often linked to but different from resistance, which is a separate word (*muqawama*). *Samoud*, in this context, is explicitly an alternative to either passivity or violent struggle; while there can be an "active" or "resistant" *samoud* (*sumud muqawama*), in its original context it translates to "simply remaining in place" or "simply carrying on" being rooted in the land, like an olive tree.

27. Dan Sallitt, "In Defense of Julia Leigh's 'Sleeping Beauty,'" *Mubi* (blog), January 16, 2012, https://mubi.com/notebook/posts/in-defense-of-julia-leighs-sleeping -beauty.

28. Julia Leigh, "Director's Notes," press kit for *Sleeping Beauty*, August 9, 2011, IFC Films.

29. Yue, "Two Sleeping Beauties," 36.

30. Lippit, *Atomic Light*, specifically 35–60.

31. Trachtenberg, "Lincoln's Smile," 12.

32. Warren and Brandeis, "The Right to Privacy."

33. Osucha, "The Whiteness of Privacy," 78.

34. Yue, "The Generic Face."

35. Atanasoski and Vora, *Surrogate Humanity*, 123, 127.

36. Osucha, "The Whiteness of Privacy," 73.

37. boyd, "White Flight in Networked Publics."

38. Chun and Friedland, "Habits of Leaking," 15; also see Hu, *A Prehistory of the Cloud*, chapter 2, for more on how the technology of virtualization creates a sanitary partition between accounts—and thus creates the idea of the user *as* private.

39. "How does a visual medium deal with interiority? How can film *show* women?" asks Eugenie Brinkema, invoking not just film theorists but also the pornographer's dilemma: How can her pleasure and ultimately her orgasm be shown? ("Celluloid Is Sticky," 153).

40. For more on the construction of middle-class interiority through portrait photography in the nineteenth century, see Smith, *American Archives*; for the way that photography turned criminals into types, see Sekula, "The Body and the Archive"; on the relationship between subjectivity and physiology, see Crary, *Techniques of the Observer*.

41. Galloway, "Black Box, Black Bloc."

42. Davis, "Imperceptibility and Accumulation," 187.

43. My use of the term builds on but differs from how Slavoj Žižek and Robert Pfaller define "interpassivity," the delegation of the work of interacting to another object or medium, which in turn allows us to be passive ourselves. (In digital culture, we increasingly imbue daemons, servers, virtual assistants, human robots, and other programs to be our proxies.) While Žižek considers interpassivity through the psychic effects on the recording device or television's "owner," he does not account for the power dynamics that adhere to the idea of ownership, nor the interpassive interaction between agents, the focus of this book (Žižek, "Will You Laugh for Me, Please?," *In These Times*, July 18, 2003, https://inthesetimes.com /article/will-you-laugh-for-me-please).

44. The exception that proves the rule: at one moment Lucy responds to a coworker with a goofy "Aye aye me hearty."

45. Chen, *Animacies*, 50.

46. Chen gives the example of the tongue-twisting sentence "the hikers that rocks crush," which is confounding to an English speaker because we normally assume that hikers have more agency than rocks—yet the rocks are the active agents in the sentence (*Animacies*, 2).

47. Chen, *Animacies*, 46.

48. For more on how "passive sex" complicates agency, see Leo Bersani, "Is the Rectum a Grave?"

49. Kim, "Why Do Dolls Die?," 95.

50. Cheng, "Ornamentalism," 442.

51. Crary, 24/7.

52. Leigh, "Sleeping Beauty," 68. Setting the actor's directions aside, further evidence for Lucy's desire to remain asleep comes from a previous exchange, where her coworker jokes about the drug he offers her bringing her "back from the dead," and Lucy wryly answers, "Fear of death is the number one hoax."

53. Fuller, *How to Sleep*, 19.

54. Figlerowicz, "Inanimism," 42.

55. Figlerowicz, "Inanimism," 49.

56. Lawrence Lek, for instance, has coined the term "Sinofuturism" to reclaim the seemingly copycat or robotic characteristics of the Chinese. See Lawrence Lek's "Sinofuturism (1839–2046 AD)" (video), and also https://medium.com /@sinethetamag/conversation-lawrence-lek-talks-sinofuturism-automation -identity-and-communism-1ce35f2a11d1.

57. Shell, *Hide and Seek*.

58. Davis, "Imperceptibility and Accumulation," 188.

59. See Blas, "Opacities: An Introduction," and the other articles in the issue Blas edited. For her part, Davis proposes a "politics of imperceptibility," one that involves "creating structures and ways of being that are at once immediately apparent, at once immediately understood, but yet reveal nothing, containing no truth, no depth" ("Imperceptibility and Accumulation," 190). For more on the difference between opacity and the common-sense meaning of it as "hiding something," see Villiers, *Opacity and the Closet*.

60. Glissant, *Poetics of Relation*, 192.

61. Amazement may seem like the opposite of lethargy, but its etymology derives from the word "maze" ("delusion, confusion"; "in a maze") and may be related to North Germanic cognates Norwegian *mas* "exhausting labor," Swedish *masa* "idle, slow, loiter." See "maze, n.1," *OED Online*, September 2021, Oxford University Press, http://www.oed.com/viewdictionaryentry/Entry/115347.

Chapter 5

1. Erica Scourti, "transmediale 2015: Erica Scourti, Banks of Body Parts and Body Scan," interview by Randy Astle, *Filmmaker*, February 24, 2015, https://filmmak

ermagazine.com/92720-transmediale-2015-erica-scourti-banks-of-body-parts-and
-body-scan/.

2. For more on "parasitism" as an artistic tactic, see Fisher, *Play in the System*.
Fisher considers artists who act as parasites to digital platforms, while Scourti arguably reverses this equation by allowing herself to be the host who is parasitized.

3. Daniel Rourke, "Artist Profile: Erica Scourti," *Rhizome*, October 8, 2013, http://
rhizome.org/editorial/2013/oct/08/artist-profile-erica-scourti/.

4. See, for example, Christian Sandvig's analysis of how algorithms cause "corrupt personalization": as he explains, "You have legitimate interests that we'll
call 'authentic.' These interests arise from your values, your community, your
work, your family, how you spend your time, and so on." Corrupt personalization,
Sandvig continues, is when algorithms feed you interests that are not "authentically" yours, but come, for example, from commercial sponsors or even hostile
state actors. Christian Sandvig, "Corrupt Personalization," Social Media Collective blog (Microsoft Research), June 26, 2014, https://socialmediacollective.org
/2014/06/26/corrupt-personalization/. For a longer history of authenticity, see
Sarah Banet-Weiser, *Authentic^TM*.

5. Scourti, personal communication, London, August 1, 2018.

6. Trilling, *Sincerity and Authenticity*.

7. See also Magill, *Sincerity*.

8. As Chun argues, algorithmic recommender systems push an idea of authenticity as transgression or deviation to predict user behavior and also to further
entrench users inside (profitable) bubbles of polarization. *Discriminating Data*, 161.

9. Turner, *From Counterculture to Cyberculture*, 2.

10. Barlow, "Declaration of the Independence of Cyberspace."

11. Scourti, "Presentation by Erica Scourti—Expose and Repurpose," YouTube
video, 20 min. 44 sec., March 18, 2015, https://www.youtube.com/watch?v=sYPd
-CTwrzA.

12. Scourti, "transmediale 2015."

13. Gomez-Uribe and Hunt, "The Netflix Recommender System."

14. Amatriain, "Big & Personal," 2.

15. Wendy Chun also describes how new media is built on the "embrace of singular yet plural individuality" ("Big Data as Drama," 364).

16. See Koopman, *How We Became Our Data.*

17. Deleuze, "Postscript on the Societies of Control," 5.

18. Chun, *Discriminating Data*, 81–120.

19. Dean, "Rich Meme, Poor Meme."

20. See Parks, "Water, Energy, Access." Parks redefines Internet infrastructure to include a broader media infrastructure around it, such as the water carriers that sustain the people and ultimately make possible the rural schoolhouses where the Internet can be accessed in Zambia.

21. National Institute of Standards and Technology, "NIST Study Evaluates Effects of Race, Age, Sex on Face Recognition Software," December 19, 2019, https://www.nist.gov/news-events/news/2019/12/nist-study-evaluates-effects-race-age-sex-face-recognition-software.

22. See also Browne, *Dark Matters*, 162.

23. Glissant, "One World in Relation," 5.

24. As Zeynep Tufekci puts it, a better Internet could in theory "help people reveal their (otherwise private) preferences to one another and discover common ground" (*Twitter and Tear Gas*, 26). Also see Fred Turner's description of the Internet's utopian ideal as one where "the individual self, so long trapped in the human body, would finally be free to step outside its fleshy confines, explore its authentic interests, and find others with whom it might achieve communion" (*From Counterculture to Cyberculture*, 1).

25. Cohen, *Never Alone*, 116. As he points out, while other mass media, such as movies, books, or television, have long obscured the identities of fellow strangers who also constitute the public for it, the long tail of a website or app can mean that possibly nobody else is "there" interacting with media at the same time.

26. Turkle, *Alone Together.*

27. The Instagram account @insta_repeat collects many of these repetitions.

28. I take the word "group form" from Cohen's *Never Alone.*

29. Cohen, *Never Alone*, 115.

30. See Steyerl, "In Defense of the Poor Image," and Althoff, "Seeping Out."

31. Erica Scourti, "Think You Know Me," YouTube video, 7 min. 2 sec., July 10, 2015, https://www.youtube.com/watch?v=qlB9l1fC3L4.

32. HEK (Haus der Elektronischen Künste), "Poetics and Politics of Data: Erica Scourti—Think You Know Me," Vimeo video, 6 min. 53 sec., July 15, 2015, https://vimeo.com/133554414.

33. Jones, "Glitch Poetics," 245.

34. As Jessica Pressman has shown, techniques such as enjambment are very much present in newer mediums; describing Young-Hae Chang Heavy Industry's Flash video poem "Dakota," she gives the example of the ambiguity that's formed when one screen of words in the poem is linked to the words on the screen that replaces it, or when a subtle delay in presenting words causes increasing moments of lexical ambiguity: about YHCHI's "I CRIED—TO THE GUYS—TO GET SMASHED," Pressman asks rhetorically, "Is the narrator crying or yelling, expressing vulnerability or evading it through drunkenness?" ("The Strategy of Digital Modernism," 311).

35. Levin, "Toward a Social Cinema Revisited," 34. The "stuff" of CNN footage that serves as Fast's raw material represents the viewers there, Levin argues, in the negative, as a "mass" of stuff governed by parataxis.

36. Terranova, *Network Culture*, 138.

37. Terranova, *Network Culture*, 138.

38. Petro, "Mass Culture and the Feminine," 6; Parks, "Flexible Microcasting."

39. Vasey, "Profile: Erica Scourti," 20.

40. Erica Scourti, "Woman Nature Alone," YouTube playlist (203 videos), December 9, 2013, https://www.youtube.com/playlist?list=PL9184D572EDB0E02B.

41. As Jean Baudrillard puts it, these tears reflect what might happen if the masses embraced their supposedly nonsubjective qualities—as insipid, infantile, and hyperconforming to the "idea" of emotion—and yet they are more potent because of their blankness (*In the Shadow of the Silent Majorities*).

42. Williams, "Film Bodies."

43. Gregg, "Inside the Data Spectacle," 45.

44. Adam Kirsch, "Obama Bests Clinton at Craft of Writing," *(New York) Sun*, March 3, 2008, https://www.nysun.com/arts/obama-bests-clinton-at-craft-of-writing /72169/.

45. Herbert Ihering, "UFA und Buster Keaton" (1926), in *Von Reinhardt bis Brecht*, vol. 2, 1924–1929 (Berlin: Aufbau Verlag, 1961), 59, as quoted by Petro, "Mass Culture and the Feminine," 9.

46. Kevin Reeve, "Gray Man Strategies 101: Peeling away the Thin Veneer of Society," *Imminent Threat Solutions* (blog), November 22, 2013, https://www .itstactical.com/intellicom/mindset/gray-man-strategies-101-peeling-away-the-thin -veneer-of-society/.

47. Further, to assume that there is some sort of "normal" behavior in the first place is a misunderstanding of how digital culture works, which now profits from difference and "authenticity." For more on TrackMeNot and other strategies of obfuscation, see Brunton and Nissenbaum, *Obfuscation*.

48. This point is elegantly made by Kate Crawford, "The Anxieties of Big Data."

49. Selvaggio, "URME Surveillance," 183.

50. Griffin, *Feeling Normal*, 1.

51. Charles Duhigg, "How Companies Learn Your Secrets," *New York Times Magazine*, February 16, 2012, https://www.nytimes.com/2012/02/19/magazine /shopping-habits.html.

52. This point is made in more detail by technology researcher Sara Watson, who describes the phenomenon of "uncanny personalization" in "Data Doppelgängers and the Uncanny Valley of Personalization," *The Atlantic*, June 16, 2014, https:// www.theatlantic.com/technology/archive/2014/06/data-doppelgangers-and-the -uncanny-valley-of-personalization/372780/.

53. Digital Methods Summer School (University of Amsterdam), "Containing Homophily," August 2, 2018, https://wiki.digitalmethods.net/Dmi/SummerSchool 2018ContainingHomophily.

54. Chun, *Discriminating Data*, 85; for the point about the emoticon, see Michael Hornsby, "Five Ways to Break Out of Your Echo Chamber," *High Snobiety*, June 27, 2017, https://www.highsnobiety.com/p/echo-chamber-fix/.

55. Note that emotional capitalism has never been simply an exploitative practice, as Eva Illouz points out. Feminists in the 1970s hoped that the same focus on

getting men to communicate in the workplace might also promote gender equality in the home (*Cold Intimacies*).

56. Illouz, *Cold Intimacies*, 108.

57. For more on normalization, see Gertenbach and Mönkeberg, "Lifelogging and Vital Normalism."

58. See the introduction, note 49.

59. Ahmed, "Melancholic Universalism" (blog entry), December 15, 2015, https://feministkilljoys.com/2015/12/15/melancholic-universalism/.

60. Cheney-Lippold, *We Are Data*.

61. In technical language, it turns something that is digital—that is, page views arriving at discrete times that happen to be roughly coincident—into an analog measure—that is, simultaneity, as in viewers of a TV broadcast.

Chapter 6

1. Deleuze, "Postscript on Societies of Control," 7.

2. Kiva Systems, "The Nutcracker Performed by Dancing Kiva Order Fulfillment Robots," YouTube video, 1 min. 24 sec., December 7, 2007, https://www.youtube.com/watch?v=Vdmtya8emMw.

3. Sekula, "Between the Net," 28–29.

4. For more on how to read the infrastructure of New York, see Ingrid Burrington, "Seeing Networks in New York City," http://seeingnetworks.in/nyc/, as well as her book, *Networks of New York*.

5. nibia pastrana santiago, *el weather bureau* (blog), February 20, 2017, https://elweatherbureau.tumblr.com/.

6. Ruiz, *Ricanness*, 9.

7. pastrana, "Datos sobre la bahía," in *maniobra*, n.p.

8. Galloway, *Protocol*, 122.

9. pastrana, "Datos sobre la bahía."

10. pastrana, *objetos indispuestos*, 41–42.

11. Pérez Varela, "Puerto Rico en la agenda tecnológica," 71.

12. Schwartz, *Sea of Storms*, 143, 188.

13. Schwartz, *Sea of Storms*, 192.

14. Schwartz, *Sea of Storms*, 217.

15. Starosielski, *The Undersea Network*, 107–108.

16. Robin James, "Yo: It's Communicative Capitalism" (blog), June 26, 2014, https://thesocietypages.org/cyborgology/2014/06/26/yo-its-communicative-capitalism/.

17. Peters, *The Marvelous Clouds*, 14.

18. pastrana, "Datos sobre la bahía."

19. Harney and Moten, *The Undercommons*, specifically 84–99.

20. Larisa Yarovaya and Brian Lucey, "Bitcoin Rich Kids in Puerto Rico: Crypto Utopia or Crypto-Colonialism?," *The Conversation*, February 14, 2018, https://theconversation.com/bitcoin-rich-kids-in-puerto-rico-crypto-utopia-or-crypto-colonialism-91527.

21. Christian Reeves, "Puerto Rico's Where It's at for Crypto Investors," *Escape Artist* (blog), February 9, 2018, https://www.escapeartist.com/blog/puerto-rico-crypto-investors/.

22. Nellie Bowles, "Making a Crypto Utopia in Puerto Rico," *New York Times*, February 2, 2018, https://www.nytimes.com/2018/02/02/technology/cryptocurrency-puerto-rico.html.

23. Priewe, "The Commuting Island," 138.

24. For an example of blockading as a tactic, see Jasper Bernes, "Logistics, Counterlogistics, and the Communist Prospect," in *Endnotes* 3, September 2013, https://endnotes.org.uk/issues/3/en/jasper-bernes-logistics-counterlogistics-and-the-communist-prospect; for an example of the electronic equivalent, see Electronic Disturbance Theater, FloodNet, 1998, archived at https://anthology.rhizome.org/floodnet.

25. nibia pastrana santiago, "inactivity at the ports," *el evento coreográfico* (blog), March 10, 2016, https://www.nibiapastrana.com/eleventocoreografico/inactivity-at-the-ports.

26. Yarimar Bonilla and Rafael A. Boglio Martínez, "Puerto Rico in Crisis: Government Workers Battle Neoliberal Reform," NACLA (North American Congress on Latin America), January 5, 2010, https://nacla.org/article/puerto-rico-crisis-government-workers-battle-neoliberal-reform; Hugo J. Delgado-Martí, "Puerto Rico's One-Sided Class War," *Jacobin*, September 13, 2016, https://www.jacobinmag.com/2016/09/puerto-rico-debt-promesa-oversight-obama-crisis.

27. Trigo, "Anemia and Vampires," 110–111.

28. Albert G. Robinson, *The Porto Rico of Today* (1899), as quoted by Santiago-Valles, *Subject People and Colonial Discourses*, 19.

29. pastrana, *objetos indispuestos*, 6.

30. Rosario's description recalls Sandra Ruiz's argument in *Ricanness* that Ricanness is about having simultaneously a debilitating excess of time and having no time left.

31. Brathwaite, *ConVERSations with Nathaniel Mackey*, 34.

32. Pugh, "Island Movements," 17.

33. François, *Open Secrets*.

34. François, *Open Secrets*, 96.

35. Mabel Rodríguez Centeno, "Las perezas insulares," *80 Grados*, November 18, 2011, http://www.80grados.net/las-perezas-insulares/.

36. Briggs, *Reproducing Empire*, 10.

37. Bonilla, *Non-Sovereign Futures*.

38. López Rivera was convicted for conspiracy for allegedly attempting to bomb targets in the United States; his sentence has since been commuted by President Obama.

39. Fanon, *Wretched of the Earth*, 42.

40. Brian Seibert, "Dance Review: Childbirth, a Lit Torch and a Trot," *New York Times*, June 3, 2014, https://www.nytimes.com/2014/06/04/arts/dance/solos-by-dd-dorvilliers-colleagues-at-st-marks-church.html.

41. Adriana Garriga-López, "Azúcar dura y melaza vaga," *80 Grados*, June 27, 2014, http://www.80grados.net/azucar-dura-y-melaza-vaga/.

42. Compare the state of "being nothing" to Sandra Ruiz's description of Ricanness to an "unwanted being" that is already dead. For Ruiz, the question is: "how do the 'dead' live?" (*Ricanness*, 24).

43. Moten, *Poetics of the Undercommons*, 15–16.

44. Harney and Moten, *The Undercommons*, 94. Moten's take diverges, however, from Wilderson's; he moves away from Blackness as an ontological position and instead understands Blackness, as Jack Halberstam puts it in his introduction to the book, "as the willingness to be in the space that has been abandoned by colonialism" (8).

45. Harney and Moten, *The Undercommons*, 96.

46. Silverman, "The Author as Receiver," 21.

47. Silverman, "The Author as Receiver," 12.

48. Joyner indicated to me that this role is analogous to the witnesser in the dance/therapeutic practice of Authentic Movement, the tradition that informed their own collaboration. Authentic Movement, developed by dancer and therapist Mary Starks Whitehead, typically pairs two dancers who alternate between the roles of witnesser and mover.

49. Lacey, *Listening Publics*, 16.

50. Lacey, *Listening Publics*, 16.

51. pastrana, "the lazy dancer," 2013, https://www.nibiapastrana.com/lazymanifesto.

52. Yarimar Bonilla, "For Investors, Puerto Rico Is a Fantasy Blank Slate," *The Nation*, February 28, 2018, https://www.thenation.com/article/for-investors-puerto-rico-is-a-fantasy-blank-slate/.

53. Bonilla, "For Investors."

54. Ricardo Rosselló, as quoted by Bonilla, "For Investors."

Postscript

1. Tadiar, *Things Fall Away*, 60–61.

2. I use the word "coolie" in an attempt to reclaim it without forgetting the painful history that it indexes. Although the word refers to low-wage workers from both

Southeast Asia and East Asia, I grew up hearing the word pronounced "kǔ lì" in Mandarin ("bitter labor"). For me, the word recalls the legal codes that regulated the Chinese in America, codes used to both construct a pool of labor for industrialists to exploit, and also to exclude those workers from Americanness. I hope to honor such workers here even as I acknowledge that the word itself may remain ugly and offensive to many.

3. Monteiro and Nadège, "Manifesto do Aço à Pele."

BIBLIOGRAPHY

Agamben, Giorgio. "Bartleby, or On Contingency." In *Potentialities: Collected Essays in Philosophy*, translated by Daniel Heller-Roazen, 243–274. Stanford, CA: Stanford University Press, 1999.

Agamben, Giorgio. "The Noonday Demon." In *Stanzas: Word and Phantasm in Western Culture*, 3–10. Minneapolis: University of Minnesota Press, 1993.

Ahmed, Sara. *The Promise of Happiness*. Durham, NC: Duke University Press, 2010.

Alexander, Neta. "Rage against the Machine: Buffering, Noise, and Perpetual Anxiety in the Age of Connected Viewing." *Cinema Journal* 56, no. 2 (2017): 1–24.

Alkhatib, Ali, Michael S. Bernstein, and Margaret Levi. "Examining Crowd Work and Gig Work through the Historical Lens of Piecework." In *Proceedings of the 2017 CHI Conference on Human Factors in Computing Systems*, 4599–4616. New York: Association for Computing Machinery, 2017. https://doi.org/10.1145/3025453 .3025974.

Althoff, Sebastian. "Seeping Out: The Diminishment of the Subject in Hito Steyerl's *How Not to Be Seen.*" *Performance Research* 24, no. 7 (2019): 92–98.

Amatriain, Xavier. "Big & Personal: Data and Models behind Netflix Recommendations." In *ACM BigMine '13*, 1–6. New York: Association for Computing Machinery, 2013. https://doi.org/10.1145/2501221.

Arora, Payal. "Bottom of the Data Pyramid: Big Data and the Global South." *International Journal of Communication* 10 (2016): 1681–1699.

Atanasoski, Neda, and Kalindi Vora. *Surrogate Humanity: Race, Robots, and the Politics of Technological Futures.* Durham, NC: Duke University Press, 2019.

Banet-Weiser, Sarah. *Authentic™: The Politics of Ambivalence in a Brand Culture.* New York: NYU Press, 2012.

Barlow, John Perry. "A Declaration of the Independence of Cyberspace." Electronic Freedom Foundation, February 8, 1996. https://www.eff.org/cyberspace -independence.

Baudrillard, Jean. *In the Shadow of the Silent Majorities, or, The End of the Social and Other Essays.* New York: Semiotext(e), 1983.

Behar, Katherine. *Bigger than You: Big Data and Obesity.* Brooklyn, NY: Punctum Books, 2016.

Behar, Katherine. "Facing Necrophilia, or 'Botox Ethics.'" In *Object Oriented Feminism*, edited by Katherine Behar, 123–143. Minneapolis: University of Minnesota Press, 2016.

Behar, Katherine. "Interview with Katherine Behar: Do Nothing, Say Nothing." Interview by Tung-Hui Hu. In *Katherine Behar: Data's Entry/Vera Girişi*, 106–112. Istanbul: Pera Museum, 2016.

Beller, Jonathan. *The Cinematic Mode of Production: Attention Economy and the Society of the Spectacle.* Hanover, NH: Dartmouth College Press/University Press of New England, 2006.

Benjamin, Ruha. *Race after Technology: Abolitionist Tools for the New Jim Code.* Medford, MA: Polity Books, 2019.

Bennett, Jane. *Vibrant Matter.* Durham, NC: Duke University Press, 2010.

Berardi, Franco "Bifo." *After the Future.* Edited by Gary Genosko and Nicholas Thoburn. Chico, CA: AK Press, 2011.

Berlant, Lauren. "The Commons: Infrastructures for Troubling Times." *Environment and Planning D: Society and Space* 34, no. 3 (2016): 393–419.

Berlant, Lauren. *Cruel Optimism.* Durham, NC: Duke University Press, 2011.

Berlant, Lauren. "Structures of Unfeeling: *Mysterious Skin.*" *International Journal of Politics, Culture, and Society* 28 (2015): 191–213. https://doi.org/10.1007/s10767-014-9190-y.

Bersani, Leo. "Is the Rectum a Grave?" *October* 43 (1987): 197–222.

Best, Stephen. *None like Us: Blackness, Belonging, Aesthetic Life.* Durham, NC: Duke University Press, 2018.

Blackmon, Douglas. *Slavery by Another Name: The Re-Enslavement of Black Americans from the Civil War to World War II.* New York: Anchor Books, 2008.

Blas, Zach. "Opacities: An Introduction." *Camera Obscura* 31, no. 2 (2016): 149–153.

Bogost, Ian. *Alien Phenomenology, or, What It's like to Be a Thing.* Minneapolis: University of Minnesota Press, 2012.

Boltanski, Luc, and Eve Chiapello. *The New Spirit of Capitalism.* London: Verso, 2005.

Bonilla, Yarimar. *Non-Sovereign Futures: French Caribbean Politics in the Wake of Disenchantment.* Chicago: University of Chicago Press, 2015.

Bourdieu, Pierre. "Social Being, Time and the Sense of Existence." In *Pascalian Meditations,* 207–245. Stanford, CA: Stanford University Press, 2000.

Bourriaud, Nicolas. *Relational Aesthetics.* Translated by Simon Pleasance and Fronza Woods. Paris: Presses du réel, 2002.

boyd, danah. "White Flight in Networked Publics: How Race and Class Shaped American Teen Engagement with MySpace and Facebook." In *Race after the Internet,* edited by Lisa Nakamura and Peter A. Chow-White, 203–222. Abingdon: Routledge, 2011.

Brathwaite, Kamau. *ConVERSations with Nathaniel Mackey.* Staten Island, NY: We Press, 1999.

Briggs, Laura. *Reproducing Empire: Race, Sex, Science, and U.S. Imperialism in Puerto Rico.* Oakland: University of California Press, 2003.

Brinkema, Eugenie. "Celluloid Is Sticky: Sex, Death, Materiality, Metaphysics (in Some Films by Catherine Breillat)." *Women: A Cultural Review* 17, no. 2 (2006): 147–170. https://doi.org/10.1080/09574040600795739.

Brown, Wendy. "Untimeliness and Punctuality: Critical Theory in Dark Times." In *Edgework: Critical Essays on Knowledge and Politics*, 1–16. Princeton, NJ: Princeton University Press, 2005.

Browne, Simone. *Dark Matters: On the Surveillance of Blackness*. Durham, NC: Duke University Press, 2015.

Brunton, Finn, and Helen Nissenbaum. *Obfuscation: A User's Guide for Privacy and Protest*. Cambridge, MA: MIT Press, 2015.

Buchloh, Benjamin. "Conceptual Art 1962–1969: From the Aesthetic of Administration to the Critique of Institutions." *October* 55 (1990): 105–143.

Burrington, Ingrid. *Networks of New York: An Illustrated Field Guide to Urban Internet Infrastructure*. Brooklyn, NY: Melville House, 2016.

Caduff, Carlo. "Hot Chocolate." *Critical Inquiry* 45, no. 3 (2019): 787–803.

Campt, Tina. "Black Visuality and the Practice of Refusal." *Women & Performance* 29, no. 1 (2019): 79–87.

Carroll, Amy Sara. *REMEX: Toward an Art History of the NAFTA Era*. Austin: University of Texas Press, 2017.

Casilli, Antonio. "Digital Labor Studies Go Global: Toward a Digital Decolonial Turn." *International Journal of Communication* 11 (2017): 3934–3954.

Chen, Mel Y. *Animacies: Biopolitics, Racial Mattering, and Queer Affect*. Durham, NC: Duke University Press, 2012.

Cheney-Lippold, John. *We Are Data: Algorithms and the Making of Our Digital Selves*. New York: NYU Press, 2017.

Cheng, Anne Anlin. "Ornamentalism: A Feminist Theory for the Yellow Woman." *Critical Inquiry* 44, no. 3 (2018): 415–446.

Chun, Wendy. "Big Data as Drama." *ELH* 83, no. 2 (2016): 363–382.

Chun, Wendy. "Crisis, Crisis, Crisis, or Sovereignty and Networks." *Theory, Culture & Society* 28, no. 6 (2011): 91–112.

Chun, Wendy. *Discriminating Data: Correlation, Neighborhoods, and the New Politics of Recognition*. Cambridge, MA: MIT Press, 2021.

Chun, Wendy. "Race and/as Technology, or How to Do Things to Race." In *Race after the Internet*, edited by Lisa Nakamura and Peter Chow-White, 38–60. Abingdon: Routledge, 2012.

Chun, Wendy, and Sarah Friedland. "Habits of Leaking: Of Sluts and Network Cards." *differences* 26, no. 2 (2015): 1–28. https://doi.org/10.1215/10407391-3145937.

Cohen, Kris. *Never Alone, Except for Now: Art, Networks, Populations.* Durham, NC: Duke University Press, 2017.

Coleman, Beth. "Race as Technology." *Camera Obscura* 24, no. 1 (2009): 177–207.

Coleman, Rebecca. "Austerity Futures: Debt, Temporality and (Hopeful) Pessimism as an Austerity Mood." *New Formations* 87 (2016): 83–101.

Crary, Jonathan. *Techniques of the Observer.* Cambridge, MA: MIT Press, 1992.

Crary, Jonathan. *24/7: Late Capitalism and the Ends of Sleep.* London: Verso, 2013.

Crawford, Kate. "The Anxieties of Big Data." *The New Inquiry*, May 30, 2014. https://thenewinquiry.com/the-anxieties-of-big-data/.

Critical Art Ensemble. *Digital Resistance: Explorations in Tactical Media.* Brooklyn, NY: Autonomedia, 2001.

Cumston, Charles Greene. *An Introduction to the History of Medicine: From the Time of the Pharaohs to the End of the XVIIIth Century.* London: K. Paul, Trench, Trubner & Co, 1926.

Cvetkovich, Ann. *Depression: A Public Feeling.* Durham, NC: Duke University Press, 2012.

Davis, Heather. "Imperceptibility and Accumulation: Political Strategies of Plastic." *Camera Obscura* 31, no. 2 (2016): 187–193.

Day, Iyko. *Alien Capital: Asian Racialization and the Logic of Settler Colonial Capitalism.* Durham, NC: Duke University Press, 2016.

Dean, Aria. "Notes on Blacceleration." *e-flux*, December 2017. https://www.e-flux.com/journal/87/169402/notes-on-blacceleration/.

Dean, Aria. "Rich Meme, Poor Meme." *Real Life Magazine*, July 25, 2016. https://reallifemag.com/poor-meme-rich-meme/.

Dean, Jodi. *Blog Theory: Feedback and Capture in the Circuits of Drive.* Cambridge: Polity Press, 2010.

Deleuze, Gilles. "Bartleby; or, The Formula." In *Essays Critical and Clinical*, translated by Daniel W. Smith and Michael A. Greco, 68–90. Minneapolis: University of Minnesota Press, 1997.

Deleuze, Gilles. "Postscript on the Societies of Control." *October* 59 (1992): 3–7.

Dhar, Ravi. "Consumer Preference for a No-Choice Option." *Journal of Consumer Research* 24, no. 2 (1997): 215–231.

Dormon, James H. "Shaping the Popular Image of Post-Reconstruction American Blacks: The 'Coon Song' Phenomenon of the Gilded Age." *American Quarterly* 40, no. 4 (1988): 450–471.

Eberstadt, Nicholas. *Men without Work: America's Invisible Crisis*. West Conshohocken, PA: Templeton Press, 2016.

Edelman, Lee. *No Future: Queer Theory and the Death Drive*. Durham, NC: Duke University Press, 2004.

Ehrenberg, Alain. *The Weariness of the Self: Diagnosing the History of Depression in the Contemporary Age*. Translated by David Homel, Enrico Caouette, Jacob Homel, and Don Winkler. Montreal: McGill-Queen's University Press, 2009.

Ellison, Ralph. "An Extravagance of Laughter." In *The Collected Essays of Ralph Ellison: Revised and Updated*, edited by John Callahan, 617–662. New York: Modern Library, 2003.

Elson, Diane, and Ruth Pearson. "'Nimble Fingers Make Cheap Workers': An Analysis of Women's Employment in Third World Export Manufacturing." *Feminist Review* 7 (1981): 87–107.

Ensmenger, Nathan. "Making Programming Masculine." In *Gender Codes: Why Women Are Leaving Computing*, edited by Thomas J. Misa, 115–141. Washington, DC: IEEE Computer Society Press and Wiley, 2010.

Fanon, Frantz. *The Wretched of the Earth*. Translated by Constance Farrington. New York: Grove, 1963.

Farman, Jason. *Delayed Response: The Art of Waiting from the Ancient to the Instant World*. New Haven, CT: Yale University Press, 2018.

Fay, Jennifer. "Bankers Dream of Banking, or Against the Interpretation of Dreams." In *Deep Mediations*, edited by Karen Redrobe and Jeff Scheible, 123–142. Minneapolis: University of Minnesota Press, 2021.

Figlerowicz, Marta. "Inanimism: *Nymphomaniac, Under the Skin*, and Capitalist Late Style." *Camera Obscura* 2, no. 98 (2018): 41–67.

Fisher, Anna Watkins. "Atop the Digital Rubble." In *Katherine Behar: E-Waste*, 25–29. Lexington, KY: Tuska Center for Contemporary Art/University of Kentucky, 2014.

Fisher, Anna Watkins. *The Play in the System: The Art of Parasitical Resistance.* Durham, NC: Duke University Press, 2020.

Foster, Hal. "An Archival Impulse." *October* 110 (2004): 3–22.

François, Anne-Lise. *Open Secrets: The Literature of Uncounted Experience.* Stanford, CA: Stanford University Press, 1999.

Franklin, Seb. *The Digitally Disposed: Racial Capitalism and the Informatics of Value.* Minneapolis: University of Minnesota Press, 2021.

Frayne, David. *The Refusal of Work.* London: Zed Books, 2015.

Freeman, Elizabeth. *Time Binds: Queer Temporalities, Queer Histories.* Durham, NC: Duke University Press, 2010.

Freudenberger, Herbert J. "Staff Burn-Out." *Journal of Social Issues* 30, no. 1 (1974): 159–165.

Fuller, Matthew. *How to Sleep: The Art, Biology and Culture of Unconsciousness.* London: Bloomsbury Academic, 2018.

Galison, Peter. "The Ontology of the Enemy: Norbert Wiener and the Cybernetic Vision." *Critical Inquiry* 21 (1994): 228–266.

Galloway, Alexander. "Black Box, Black Bloc." In *Communization and Its Discontents: Contestation, Critique, and Contemporary Struggles*, edited by Benjamin Noys, 237–252. Brooklyn, NY: Autonomedia, 2011.

Galloway, Alexander. *The Interface Effect.* Malden, MA: Polity Books, 2012.

Galloway, Alexander. *Protocol: How Control Exists after Decentralization.* Cambridge, MA: MIT Press, 2004.

Galloway, Alexander, and Eugene Thacker. *The Exploit: A Theory of Networks.* Minneapolis: University of Minnesota Press, 2007.

Geissler, Heike. *Seasonal Associate.* Translated by Katy Derbyshire. South Pasadena, CA: Semiotext(e), 2018.

Gershon, Ilana. *Down and Out in the New Economy: How People Find (or Don't Find) Work Today.* Chicago: University of Chicago Press, 2017.

Gershon, Ilana. "Neoliberal Agency." *Current Anthropology* 52, no. 4 (August 2011): 537–555.

Gertenbach, Lars, and Sarah Mönkeberg. "Lifelogging and Vital Normalism: Sociological Reflections on the Cultural Impact of the Reconfiguration of Body and Self." In *Lifelogging*, edited by Stefan Selke, 25–42. Wiesbaden: Springer VS, 2016.

Giardina Papa, Elisa. "Technologies of Care." *Rhizome*, October 4, 2016. http:// rhizome.org/editorial/2016/oct/04/the-download-technologies-of-care/.

Glissant, Édouard. "Natural Poetics, Forced Poetics." In *Caribbean Discourse: Selected Essays*, translated by J. Michael Dash, 120–134. Charlottesville: University Press of Virginia, 1989.

Glissant, Édouard. "One World in Relation: Édouard Glissant in Conversation with Manthia Diawara." Interview by Manthia Diawara. *Nka* 28 (2011): 4–19.

Glissant, Édouard. *Poetics of Relation*. Translated by Betsy Wing. Ann Arbor: University of Michigan Press, 1990.

Goldberg, Greg. *Anti-Social Media: Anxious Labor in the Digital Economy*. New York: NYU Press, 2018.

Goldsmith, Kenneth. *Uncreative Writing: Managing Language in the Digital Age*. New York: Columbia University Press, 2011.

Gomez-Uribe, Carlos A., and Neil Hunt. "The Netflix Recommender System: Algorithms, Business Value, and Innovation." *ACM Transactions on Management Information Systems* 6, no. 4 (December 28, 2015): 1–19. https://doi.org/10.1145 /2843948.

Gorfinkel, Elena. "Weariness, Waiting: Endurance and Art Cinema's Tired Bodies." *Discourse* 34, nos. 2–3 (Spring/Fall 2012): 311–347.

Gray, Mary, and Siddharth Suri. *Ghost Work: How to Stop Silicon Valley from Building a New Global Underclass*. Boston: Houghton Mifflin, 2019.

Green, Venus. *Race on the Line: Gender, Labor, and Technology in the Bell System, 1880–1980*. Durham, NC: Duke University Press, 2001.

Greene, Amanda. "Modern Feels: Interwar Britain and the Bodily Politics of Visual Social Media." PhD dissertation, University of Michigan, 2019.

Gregg, Melissa. "Inside the Data Spectacle." *Television and New Media* 16, no. 1 (2015): 37–51.

Griffin, F. Hollis. *Feeling Normal: Sexuality and Media Criticism in the Digital Age.* Bloomington: University of Indiana Press, 2016.

Halberstam, Jack. *The Queer Art of Failure.* Durham, NC: Duke University Press, 2011.

Han, Byung-Chul. *The Burnout Society.* Stanford, CA: Stanford University Press, 2010.

Harman, Graham. *The Quadruple Object.* Winchester: Zero Books, 2011.

Harney, Stefano, and Fred Moten. *The Undercommons: Fugitive Planning & Black Study.* Wivenhoe: Minor Compositions, 2013.

Hartman, Saidiya. *Wayward Lives, Beautiful Experiments.* New York: Norton, 2019.

Heeks, Richard. *Decent Work and the Digital Gig Economy: Development Informatics Working Paper No. 71.* Centre for Development Informatics, University of Manchester, 2017. https://hummedia.manchester.ac.uk/institutes/gdi/publications/workingpapers/di/di_wp71.pdf.

Hefty, Adam. "Labor and Lamentation: A Genealogy of Acedia, Alienated Labor, and Depressed Affects." PhD dissertation, University of California, Santa Cruz, 2013.

Heller-Roazen, Daniel. *The Enemy of All: Piracy and the Law of Nations.* New York: Zone Books, 2009.

Hitlin, Paul, and Lee Rainie. "Facebook Algorithms and Personal Data." *Pew Research Center,* January 16, 2019. https://www.pewresearch.org/internet/wp-content/uploads/sites/9/2019/01/PI_2019.01.16_Facebook-algorithms_FINAL2.pdf.

Hodge, James. "Sociable Media: Phatic Connection in Digital Art." *Postmodern Culture* 28, no. 1 (2015): 10. https://doi.org/10.1353/pmc.2015.0021.

Hodge, James. "Touch." *TriQuarterly* 154 (2018). http://www.triquarterly.org/node/303186.

Hong, Cathy Park. *Minor Feelings: An Asian American Reckoning.* New York: One World, 2020.

Hsu, Hsuan. *Sitting in Darkness: Mark Twain's Asia and Comparative Racialization.* New York: NYU Press, 2015.

Hu, Tung-Hui. *A Prehistory of the Cloud*. Cambridge, MA: MIT Press, 2015.

Hu, Tung-Hui. "Real Time/Zero Time." *Discourse* 34, nos. 2–3 (2012): 163–184.

Hu, Tung-Hui. "Work at the Bleeding Edge of Sovereignty." In *Former West: Art and the Contemporary after 1989*, edited by Maria Hlavajova and Simon Sheikh, 467–476. Utrecht: BAK, basis voor actuele kunst/Cambridge, MA: MIT Press, 2017.

Illouz, Eva. *Cold Intimacies: The Making of Emotional Capitalism*. Hoboken, NJ: Wiley, 2007.

Irani, Lilly. "The Cultural Work of Microwork." *New Media & Society* 17, no. 5 (2013): 720–739.

Jagoda, Patrick. *Network Aesthetics*. Chicago: University of Chicago Press, 2016.

Jennings, Humphrey. *Spare Time*. Film. GPO (General Post Office) Film Unit, 1939.

Jones, Nathan. "Glitch Poetics: The Posthumanities of Error." In *The Bloomsbury Handbook of Electronic Literature*, edited by Joseph Tabbi, 237–252. London: Bloomsbury Academic, 2017.

Joselit, David. *Feedback: Television against Democracy*. Cambridge, MA: MIT Press, 2006.

Keeling, Kara. *Queer Times, Black Futures*. New York: NYU Press, 2019.

Kevorkian, Martin. *Color Monitors: The Black Face of Technology in America*. Ithaca, NY: Cornell University Press, 2006.

Kim, Eunjung. "Why Do Dolls Die? The Power of Passivity and the Embodied Interplay between Disability and Sex Dolls." *Review of Education, Pedagogy, and Cultural Studies* 34, nos. 3–4 (2012): 94–106.

Kittler, Friedrich. *Gramophone, Film, Typewriter*. Translated by Geoffrey Winthrop-Young and Michael Wutz. Stanford, CA: Stanford University Press, 1999.

Koopman, Colin. *How We Became Our Data: A Genealogy of the Informational Person*. Chicago: University of Chicago Press, 2019.

Krajewski, Markus. *The Server: A Media History from the Present to the Baroque*. Translated by Ilinca Iurascu. New Haven, CT: Yale University Press, 2018.

Kury, Patrick. "Neurasthenia and Managerial Disease in Germany and America: Transnational Ties and National Characteristics in the Field of Exhaustion 1880–1960." In *Burnout, Fatigue, Exhaustion: An Interdisciplinary Perspective on a Modern Affliction*, edited by Sighard Neckel, Anna Katharina Schaffner, and Greta Wagner, 51–73. London: Palgrave Macmillan, 2017.

Lacey, Kate. *Listening Publics: The Politics and Experience of Listening in the Media Age*. Malden, MA: Polity Press, 2013.

Lazzarato, Maurizio. "Immaterial Labor." In *Radical Thought in Italy*, edited by Paolo Virno and Michael Hardt, translated by Paul Colilli and Ed Emory, 133–147. Minneapolis: University of Minnesota Press, 1996.

Lee, Summer Kim. "Staying In: Mitski, Ocean Vuong, and Asian American Asociality." *Social Text* 37, no. 1 (2019): 27–50.

Leigh, Julia. "Sleeping Beauty." Film shooting script, 2011. Core Collection, Scripts, Margaret Herrick Library, Academy of Motion Picture Arts and Sciences.

Lepecki, André. "Undoing the Fantasy of the (Dancing) Subject: 'Still Acts' in Jérôme Bel's *The Last Performance*." In *The Salt of the Earth: On Dance, Politics and Reality*, edited by Steven de Belder and Koen Tachelet. Brussels: Vlaams Theater Instituut, 2001. 43-54.

Levin, Erica. "Toward a Social Cinema Revisited." *Millennium Film Journal* 58 (2013): 30–36.

Levinas, Emmanuel. *Existence and Existents*. Translated by Alphonso Lingis. London: Kluwer Academic, 1995.

Lewis, Kristen. *A Decade Undone: Youth Disconnection in the Age of Coronavirus*. New York: Measure of America/Social Science Research Council, 2020.

Lindtner, Silvia. *Prototype Nation: China and the Contested Promise of Innovation*. Princeton, NJ: Princeton University Press, 2020.

Lippit, Akira. *Atomic Light (Shadow Optics)*. Minneapolis: University of Minnesota Press, 2004.

Litwack, Michael. "Making Television Live: Mediating Biopolitics in Obesity Programming." *Camera Obscura* 1, no. 88 (January 2015): 41–69.

Liu, Alan. *The Laws of Cool: Knowledge Work and the Culture of Information*. Chicago: University of Chicago Press, 2004.

Lonergan, Guthrie. "Hacking vrs. Defaults." Guthrie Lonergan blog, January 10, 2007. http://guthguth.blogspot.com/2007/01/hacking-defaults-hacking-nintendo.html.

Lowe, Lisa. *The Intimacies of Four Continents.* Durham, NC: Duke University Press, 2015.

Lütticken, Sven. "Liberation through Laziness. Some Chronopolitical Remarks." *Mousse* 42 (2014). http://moussemagazine.it/sven-luetticken-refusal-2014/.

Lunenfeld, Peter. *The Secret War between Downloading and Uploading: Tales of the Computer as Culture Machine.* Cambridge, MA: MIT Press, 2011.

Magill, R. J. *Sincerity: How a Moral Ideal Born Five Hundred Years Ago Inspired Religious Wars, Modern Art, Hipster Chic, and the Curious Notion That We All Have Something to Say (No Matter How Dull).* New York: Norton, 2012.

Mahmood, Saba. *Politics of Piety: The Islamic Revival and the Feminist Subject.* Princeton, NJ: Princeton University Press, 2004.

Mankekar, Purnima. *Unsettling India: Affect, Temporality, Transnationality.* Durham, NC: Duke University Press, 2015.

Marez, Curtis. *Farm Worker Futurism: Speculative Technologies of Resistance.* Minneapolis: University of Minnesota Press, 2016.

Mbembe, Achille. "Africa and the Future." Interview by Thomas Blaser. *Africa Is a Country*, November 20, 2013. https://africasacountry.com/2013/11/africa-and-the-future-an-interview-with-achille-mbembe/.

Mbembe, Achille. *Necropolitics.* Durham, NC: Duke University Press, 2019.

McMillan, Uri. *Embodied Avatars: Genealogies of Black Feminist Art and Performance.* New York: NYU Press, 2015.

Melamed, Jodi. "Racial Capitalism." *Critical Ethnic Studies* 1, no. 1 (2015): 76–85.

Mengesha, Lilian, and Lakshmi Padmanabhan. "Introduction: Performing Refusal/Refusing to Perform." *Women & Performance* 29, no. 1 (2019): 1–8.

Mingo, Jack. *The Official Couch Potato Handbook: A Guide to Prolonged Television Viewing.* Santa Barbara, CA: Capra Press, 1983.

Monteiro, Nanda [Fernanda], and Nadège. "Manifesto do Aço à Pele." *Kefir*, July 17, 2017. https://web.archive.org/web/20190218032847/https://fermentos.kefir.red/english/aco-pele/.

Moten, Fred. *Poetics of the Undercommons*. Brooklyn, NY: Sputnik & Fizzle, 2016.

Nakamura, Lisa. "Don't Hate the Player, Hate the Game: The Racialization of Labor in World of Warcraft." *Critical Studies in Media Communication* 26, no. 2 (June 2009): 128–144.

Ngai, Sianne. *Ugly Feelings*. Cambridge, MA: Harvard University Press, 2005.

Nunes, Mark, ed. *Error: Glitch, Noise, and Jam in New Media Cultures*. New York: Continuum, 2011.

Okoth, Kevin Ochieng. "The Flatness of Blackness: Afro-Pessimism and the Erasure of Anti-Colonial Thought." *Salvage*, January 16, 2020. https://salvage.zone /issue-seven/the-flatness-of-blackness-afro-pessimism-and-the-erasure-of-anti -colonial-thought/.

O'Neill, Cathy. *Weapons of Math Destruction*. New York: Crown, 2016.

Osucha, Eden. "The Whiteness of Privacy: Race, Media, Law." *Camera Obscura* 24, no. 1 (70) (2009): 66–107.

Parks, Lisa. "Flexible Microcasting: Gender, Generation, and Television-Internet Convergence." In *Television after TV: Essays on a Medium in Transition*, edited by Lynn Spigel and Jan Olsson, 133–156. Durham, NC: Duke University Press, 2004.

Parks, Lisa. "'Stuff You Can Kick': Toward a Theory of Media Infrastructures." In *Between Humanities and the Digital*, edited by Patrik Svensson and David Theo Goldberg, 355–373. Cambridge, MA: MIT Press, 2015.

Parks, Lisa. "Water, Energy, Access: Materializing the Internet in Rural Zambia." In *Signal Traffic: Critical Studies of Media Infrastructures*, edited by Lisa Parks and Nicole Starosielski, 115–136. Champaign: University of Illinois Press, 2016.

Parvulescu, Anca. "Even Laughter? From Laughter in the Magic Theater to the Laughter Assembly Line." *Critical Inquiry* 43, no. 2 (2017): 506–527.

pastrana santiago, nibia. *maniobra, bahía o el evento coreográfico*. Santurce: La Impresora, 2016.

pastrana santiago, nibia. *nibia pastrana santiago, objetos indispuestos, inauguraciones suspendidas o finales inevitables para un casi-baile*. Edited by David Buuck. Oakland, CA: Tripwire, 2019.

Pérez Varela, Tomás. "Puerto Rico en la agenda tecnológica de Estados Unidos 1890–1912: Telecomunicación global y colonialism." PhD dissertation, Universidad de Puerto Rico—Río Piedras, 2015.

Peters, John Durham. *The Marvelous Clouds: Toward a Philosophy of Elemental Media*. Chicago: University of Chicago Press, 2015.

Petro, Patrice. "Mass Culture and the Feminine: The 'Place' of Television in Film Studies." *Cinema Journal* 25, no. 3 (1986): 5–21.

Piao, Guangyuan, and John G. Breslin. "Inferring User Interests for Passive Users on Twitter by Leveraging Followee Biographies." In *Advances in Information Retrieval: 39th European Conference on IR Research*, 122–133. Heidelberg: Springer, 2017.

Pressman, Jessica. "The Strategy of Digital Modernism: Young-Hae Chang Heavy Industries's *Dakota*." *MFS Modern Fiction Studies* 54, no. 2 (2008): 302–326.

Priewe, Marc. "The Commuting Island: Cultural (Im)mobility in *The Flying Bus*." In *Kulturelle Mobilitätsforschung: Themen—Theorien—Tendenzen*, edited by Norbert Franz and Rüdiger Kunow, 135–148. Potsdam: Universitätsverlag Potsdam, 2011.

Pugh, Jonathan. "Island Movements: Thinking with the Archipelago." *Island Studies Journal* 8, no. 1 (2013): 9–24.

Quashie, Kevin. *The Sovereignty of Quiet: Beyond Resistance in Black Culture*. New Brunswick, NJ: Rutgers University Press, 2012.

Rabinbach, Anson. *The Human Motor: Energy, Fatigue and the Origins of Modernity*. Oakland: University of California Press, 1992.

Rajagopal, Arvind. *Politics after Television: Hindu Nationalism and the Reshaping of the Public in India*. Cambridge: Cambridge University Press, 2001.

Rakow, Lana. "Women and the Telephone: The Gendering of a Communications Technology." In *Technology and Women's Voices: Keeping in Touch*, edited by Cheris Kramarae, 207–228. New York: Routledge and Kegan Paul (Methuen), 1988.

Raley, Rita. *Tactical Media*. Minneapolis: University of Minnesota Press, 2009.

Rauch, Jennifer. *Slow Media: Why Slow Is Satisfying, Sustainable, and Smart*. Chicago: University of Chicago Press, 2018.

Rhee, Jennifer. *The Robotic Imaginary: The Human and the Price of Dehumanized Labor*. Minneapolis: University of Minnesota Press, 2018.

Richmond, Scott. "Vulgar Boredom, or What Andy Warhol Can Teach Us about *Candy Crush*." *Journal of Visual Culture* 14, no. 1 (2015): 21–39.

Roberts, Sarah T. *Behind the Screen: Content Moderation in the Shadows of Social Media*. New Haven, CT: Yale University Press, 2019.

Robinson, Cedric. *Black Marxism: The Making of the Black Radical Tradition*. Chapel Hill: University of North Carolina Press, 2000.

Roh, David S., Betsy Huang, and Greta A. Niu. "Technologizing Orientalism: An Introduction." In *Techno-Orientalism: Imagining Asia in Speculative Fiction, History, and Media*, edited by David S. Roh, Betsy Huang, and Greta A. Niu, 1–22. New Brunswick, NJ: Rutgers University Press, 2015.

Rosenberg, Daniel. "The Young and the Restless: A History of Busy Idleness." *Cabinet* 29 (2008): 77–83.

Ross, Christine. *The Aesthetics of Disengagement: Contemporary Art and Depression*. Minneapolis: University of Minnesota Press, 2006.

Ruiz, Sandra. *Ricanness: Enduring Time in Anticolonial Performance*. New York: NYU Press, 2019.

Russell, Legacy. *Glitch Feminism: A Manifesto*. London: Verso, 2020.

Salvato, Nick. *Obstruction*. Durham, NC: Duke University Press, 2016.

Santiago-Valles, Kelvin A. *Subject People and Colonial Discourses: Economic Transformation and Social Disorder in Puerto Rico, 1898–1947*. Albany: SUNY Press, 1994.

Schaffner, Anna K. *Exhaustion: A History*. New York: Columbia University Press, 2016.

Schwartz, Stuart B. *Sea of Storms: A History of Hurricanes in the Greater Caribbean from Columbus to Katrina*. Princeton, NJ: Princeton University Press, 2016.

Schwarz, Stephanie. "Waiting: Loops in Time." In *Focus: Waiting for Tear Gas 1999–2000 by Allan Sekula*. Tate Research Publication, 2016. https://www.tate.org.uk/research/publications/in-focus/waiting-for-tear-gas-allan-sekula/loops-in-time.

Scott, James C. *Weapons of the Weak: Everyday Forms of Peasant Resistance*. New Haven, CT: Yale University Press, 1985.

Sekula, Allan. "Between the Net and the Deep Blue Sea (Rethinking the Traffic in Photographs)." *October* 102 (2002): 3–34.

Sekula, Allan. "The Body and the Archive." *October* 39 (1986): 3–64.

Sekula, Allan. "Waiting for Tear Gas [White Globe to Black]." In *Allan Sekula: Performance under Working Conditions*, edited by Sabine Breitweiser, 310. Vienna: Generali Foundation, 2003.

Selvaggio, Leonardo. "URME Surveillance: Performing Privilege in the Face of Automation." *International Journal of Performance Arts and Digital Media* 11, no. 2 (2015): 165–184. https://doi.org/10.1080/14794713.2015.1086138.

Serpell, Namwali. "Sun Ra: 'I'm Everything and Nothing.'" *New York Review of Books*, July 23, 2020.

Sexton, Jared. "Afro-Pessimism: The Unclear Word." *Rhizomes* 29 (2016). https://doi.org/10.20415/rhiz/029.e02.

Shalson, Lara. *Performing Endurance: Art and Politics since 1960*. Cambridge: Cambridge University Press, 2018.

Sharma, Sarah. *In the Meantime: Temporality and Cultural Politics*. Durham, NC: Duke University Press, 2014.

Sheikh, Simon. "Circulation and Withdrawal, Part II: Withdrawal." *e-flux* 63 (March 2015). https://www.e-flux.com/journal/63/60924/circulation-and-withdrawal-part-ii-withdrawal/.

Shell, Hanna Rose. *Hide and Seek: Camouflage, Photography and the Media of Reconnaissance*. New York: Zone Books, 2012.

Siegelbaum, Sami. "Business Casual: Flexibility in Contemporary Performance Art." *Art Journal* 72, no. 3 (2013): 51–65.

Silverman, Kaja. "The Author as Receiver." *October* 96 (2001): 17–34.

Sivakorn, S., I. Polakis, and A. D. Keromytis. "I Am Robot: (Deep) Learning to Break Semantic Image CAPTCHAs." In *2016 IEEE European Symposium on Security and Privacy (EuroS&P)*, 388–403. Saarbrucken, 2016. https://doi.org/10.1109/EuroSP.2016.37.

Smith, Shawn Michelle. *American Archives: Gender, Race, and Class in Visual Culture*. Princeton, NJ: Princeton University Press, 1999.

Smythe, Dallas. "Communications: Blindspot of Economics." *Canadian Journal of Political and Social Theory/Revue Canadienne de Theorie Politique et Sociale* 1, no. 3 (1977): 1–27.

Snider, Laureen. "Crimes against Capital: Discovering Theft of Time." *Social Justice* 28, no. 3 (85) (Fall 2001): 105–120.

Spillers, Hortense J. "Mama's Baby, Papa's Maybe: An American Grammar Book." *Diacritics* 17, no. 2 (1987): 64–81.

Starosielski, Nicole. *The Undersea Network.* Durham, NC: Duke University Press, 2015.

Sterne, Jonathan. "What If Interactivity Is the New Passivity?" *Flow: A Critical Forum on Media and Culture,* April 9, 2012. https://www.flowjournal.org/2012/04/the-new-passivity/.

Stewart, Kathleen. *Ordinary Affects.* Durham, NC: Duke University Press, 2007.

Steyerl, Hito. "In Defense of the Poor Image." *e-flux* 10 (November 2009). https://www.e-flux.com/journal/10/61362/in-defense-of-the-poor-image/.

Sullivan, Garrett, Jr. *Memory and Forgetting in English Renaissance Drama.* Cambridge: Cambridge University Press, 2005.

Sundén, Jenny. "On Trans-, Glitch, and Gender as Machinery of Failure." *First Monday* 20, no. 4–6 (2015).

Tadiar, Neferti X. M. "By the Waysides, or, Bypass and Splendor." *Modernism/Modernity* 2, no. 4 (January 2, 2018). https://doi.org/10.26597/mod.0036.

Tadiar, Neferti X. M. *Things Fall Away: Philippine Historical Experience and the Makings of Globalization.* Durham, NC: Duke University Press, 2009.

Terranova, Tiziana. *Network Culture: Politics for the Information Age.* London: Pluto Press, 2004.

Thompson, E. P. "Work-Discipline, and Industrial Capitalism." *Past & Present* 38 (December 1967): 56–97.

Trachtenberg, Alan. "Lincoln's Smile: Ambiguities of the Face in Photography." *Social Research* 67, no. 1 (2000): 1–23.

Trigo, Benigno. "Anemia and Vampires: Figures to Govern the Colony, Puerto Rico, 1880 to 1904." *Comparative Studies in Society and History* 41, no. 1 (January 1999): 104–123.

Trilling, Lionel. *Sincerity and Authenticity.* Cambridge, MA: Harvard University Press, 1972.

Tsing, Anna Lowenhaupt. *Friction: An Ethnography of Global Connection.* Princeton, NJ: Princeton University Press, 2004.

Tufekci, Zeynep. *Twitter and Tear Gas: The Power and Fragility of Networked Protest.* New Haven, CT: Yale University Press, 2017.

Turing, Alan M. "Computing Machinery and Intelligence." *Mind: A Quarterly Review of Psychology and Philosophy* 59, no. 236 (October 1950): 433–460.

Turkle, Sherry. *Alone Together: Why We Expect More from Technology and Less from Each Other.* New York: Basic Books, 2011.

Turner, Fred. *From Counterculture to Cyberculture: Stewart Brand, the Whole Earth Network, and the Rise of Digital Utopianism.* Chicago: University of Chicago Press, 2008.

Ueno, Toshiya. "Japanimation and Techno-Orientalism." In *The Uncanny: Experiments in Cyborg Culture,* edited by Bruce Grenville. Vancouver: Arsenal Pulp Press, 2002.

Vasey, George. "Profile: Erica Scourti." *Art Monthly* 382 (January 2014): 20–21.

Villiers, Nicholas de. *Opacity and the Closet: Queer Tactics in Foucault, Barthes, and Warhol.* Minneapolis: University of Minnesota Press, 2012.

Wall-Romana, Christophe. "Cinepoetry: Unmaking and Remaking the Poem in the Age of Cinema." PhD dissertation, University of California, Berkeley, 2005.

Wang, Jackie. *Carceral Capitalism.* South Pasadena, CA: Semiotext(e), 2018.

Warren, Samuel D., and Louis D. Brandeis. "The Right to Privacy." *Harvard Law Review* 4, no. 5 (1890): 193–220.

Weeks, Kathi. *The Problem with Work: Feminism, Marxism, Antiwork Politics, and Postwork Imaginaries.* Durham, NC: Duke University Press, 2011.

Weheliye, Alexander. *Habeas Viscus: Racializing Assemblages, Biopolitics, and Black Feminist Theories of the Human.* Durham, NC: Duke University Press, 2014.

Wilke, Hannah. "Intercourse with . . ." Performance/lecture, London Art Gallery, London, Ontario, Canada, 17 February 1977.

Williams, Linda. "Film Bodies: Gender, Genre, and Excess." *Film Quarterly* 44, no. 4 (1991): 2–13.

Wynter, Sylvia. "Beyond the Word of Man: Glissant and the New Discourse of the Antilles." *World Literature Today* 63, no. 4 (1989): 637–648.

Wynter, Sylvia. "Unsettling the Coloniality of Being/Power/Truth/Freedom: Towards the Human, after Man, Its Overrepresentation—An Argument." *CR: The New Centennial Review* 3, no. 3 (2003): 257–337.

Yue, Genevieve. "The Generic Face: Galton, Muybridge, and the Photographic Proof of Race." *Faces on Screen: New Approaches*, edited by Alice Maurice, 15–29. Edinburgh: Edinburgh University Press, 2022.

Yue, Genevieve. "Two Sleeping Beauties." *Film Quarterly* 65, no. 3 (2012): 33–37.

Zieger, Susan. "'Shipped.' Paper, Print, and the Atlantic Slave Trade." In *Assembly Codes: The Logistics of Media*, edited by Matthew Hockenberry, Nicole Starosielski, and Susan Zieger, 34–53. Durham, NC: Duke University Press, 2021.

Zorzanelli, Rafaela Teixeira. "Fatigue and Its Disturbances: Conditions of Possibility and the Rise and Fall of Twentieth-Century Neurasthenia." *História, Ciências, Saúde-Manguinhos* 16, no. 3 (July 2009). https://doi.org/10.1590/S0104-59702009000300002.

INDEX

Note: Page numbers in italics refer to images.